EXERCISES IN PRACTICAL PHYSICS

EXERCISES
IN
PRACTICAL PHYSICS

by

SIR ARTHUR SCHUSTER

PH.D., SC.D., F.R.S.

HONORARY PROFESSOR OF PHYSICS IN THE
UNIVERSITY OF MANCHESTER

and

CHARLES H. LEES

D.SC., F.R.S.

PROFESSOR OF PHYSICS IN THE UNIVERSITY OF LONDON
(EAST LONDON COLLEGE)

FIFTH EDITION
REVISED

CAMBRIDGE
AT THE UNIVERSITY PRESS
1925

CAMBRIDGE
UNIVERSITY PRESS

University Printing House, Cambridge CB2 8BS, United Kingdom

Cambridge University Press is part of the University of Cambridge.

It furthers the University's mission by disseminating knowledge in the pursuit of education, learning and research at the highest international levels of excellence.

www.cambridge.org
Information on this title: www.cambridge.org/9781107559875

First edition 1901
Second edition (revised) 1905
Third edition 1911
Fourth edition (revised) 1915
Fifth edition (revised) 1925
First paperback edition 2015

A catalogue record for this publication is available from the British Library

ISBN 978-1-107-55987-5 Paperback

PREFACE

This volume is intended for students who, having obtained an elementary knowledge of experimental work in Physics, desire to become acquainted with the principles and methods of accurate measurement. The large and increasing number of students, who have to be taught simultaneously in a Physical Laboratory, renders it necessary that the instructions supplied should be fairly complete; and that the exercises should be of such a nature as to enable the teachers easily to check the accuracy of the results obtained. The exercises described in this volume have been worked through by several hundred students of this University who were preparing for the ordinary degree of B.Sc., and the experience thus gained has been utilised to improve the descriptions and methods adopted. It is hoped therefore that the volume will also prove useful in other laboratories.

We have not aimed at completeness, being convinced that a student learns more by carefully working through a few selected and typical exercises, than by hurrying through a large number, which are often but slight modifications of each other.

The guiding principle we have adopted in our teaching has been to attach greater importance to neat and accurate work, properly recorded, than to the number of experiments which a student performs. All Note-books are carefully kept, no slovenly work is allowed to pass, and each exercise is repeated until satisfactory results have been obtained.

A student will naturally devote the greater portion of the time spent in the laboratory to measurements and quantitative work, but qualitative experiments should not be excluded. In certain parts of the subject, as for instance in Physical Optics, the educational value of setting up the apparatus and observing the general character of the effects produced is considerable, and such observations form a very useful complement to the quantitative exercises given in this book.

We have endeavoured to confine the apparatus required to that commonly found in laboratories. It is not necessary that the instruments used should be identical with those described. Students should be able to introduce the slight modifications in the manipulation rendered necessary by some small differences in the apparatus. Where the differences are likely to be material, detailed descriptions have usually been omitted, and in such cases a written explanation should be supplied to the student with the instrument to be used.

ARTHUR SCHUSTER
CHARLES H. LEES

MANCHESTER
August, 1905

PREFACE TO FIFTH EDITION

We have taken the opportunity provided by the reprinting of this book to substitute for some of the simpler exercises, which since the first edition have become of Intermediate rather than Pass Degree standard, an equal number requiring more skill on the part of the student.

A. S.
C. H. L.

CONTENTS

BOOK I

PRELIMINARY

BOOK II

GENERAL PHYSICS

BOOK III

HEAT

BOOK IV

SOUND

BOOK V

LIGHT

BOOK VI

MAGNETISM AND ELECTRICITY

BOOK I

PRELIMINARY

SECTION I

TREATMENT OF OBSERVATIONS

Errors of Observation.

OUR senses and judgment may be trusted up to certain limits, beyond which they begin to be subject to errors. Thus if we wish to measure a length of say 5 centimetres by means of a millimetre scale, no one will feel any difficulty in obtaining a result accurate to a millimetre or to half that quantity. But as soon as we wish to push the accuracy much further, even the most experienced observer will find the estimation difficult, and his measurement may be wrong by a quantity which is called an "error of observation." If he repeats the observations a great many times he will obtain a number of different results, which will group themselves round their average or mean value in a manner which will always shew a certain regularity, if the number of observations is sufficiently large. The study of the law of distribution of errors is of importance because it allows us to form an estimate of the accuracy with which under given conditions the measurements can be made. If there is no bias, which will cause the observation to err more often in one direction than in the other, common sense is sufficient to tell us that the arithmetical mean of a number of observations will give us the most probable result. And common sense will also allow us to form a rough estimate as to the limits within which the result may be trusted to be right. Suppose for instance a certain observation three times repeated has given the numbers 3·1, 3·3, 3·4, and in another case the three observations have been 1·1, 1·5, 7·2. In both instances the most probable value, being the arithmetical mean, is the same, viz. 3·27, but in the first case the observer may conclude with some confidence that his result is right to within ten per cent., *i.e.* the actual value will lie between 3 and 3·5, while in the second case he will attach little value to the mean obtained from such discordant measurements.

Common sense like the sense of sight or of hearing may be trusted up to certain limits, and just as we can increase the efficiency of our ordinary senses by suitable instruments, so may we increase the efficiency of this common sense by an instrument which in this case is the theory of probability. To apply that theory we must in the first instance study the laws of grouping of errors, and this is best done by a graphical method. Let the curved line in Fig. 1 have the property that

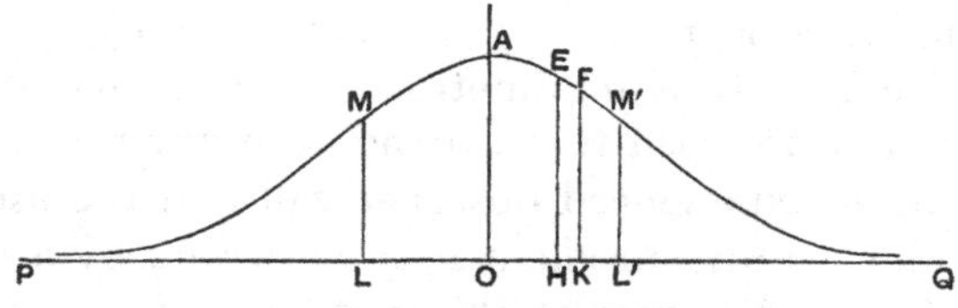

FIG. 1. Normal Probability Curve

if N represents the number of observations supposed to be very large, the area $EFHK$ will be a measure of the ratio n/N, where n is the number of observations which shew an error greater than OH and smaller than OK. It follows of course that the unit of length chosen is such that the total area included between the curve and the line PQ is unity.

It is found that in all cases which it is necessary to consider here, the curve has the same shape and may be represented analytically by the equation

$$y = \frac{h}{\sqrt{\pi}} e^{-h^2x^2},$$

where x is the "error," *i.e.* the deviation of an observation from the arithmetical mean. It is seen that different cases can only differ owing to a difference in the value of h, and Fig. 2 gives the curves for three different values of h. Inspection of these curves shews that the greater the value of

h, the steeper the curve in the neighbourhood of the central ordinate OA, and this means that the observations are grouped more closely round their average value. We might therefore take h to be the measure of the precision of our observations, but it is usual to choose for this purpose another quantity which we proceed to define. In Fig. 1 draw two lines LM and $L'M'$ at equal distances from OA such that the area included between these lines, the curve and the horizontal axis is equal to half unit area. This secures that half the total observations have errors numerically smaller than OL. The quantity OL is called the "mean" error, since

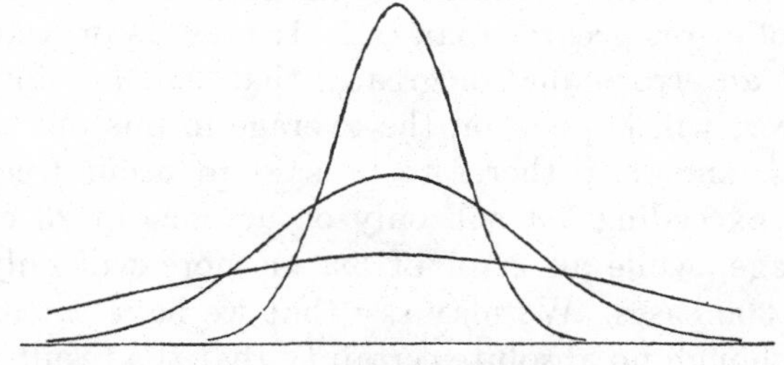

FIG. 2. Normal Probability Curves

errors larger and smaller than that quantity occur equally often. The mean error m may be calculated in terms of h from the equation to the curves and is found to be given by

$$hm = \cdot 4769.$$ (Dale's *Mathematical Tables*, p. 84.)

The mean error varies therefore inversely as h, and the smaller the mean error the more confidence may we have in our result.

The quantity which interests us most, however, is not the mean error of an observation but the mean error of the result which is obtained by taking the mean of all the observations. Assuming the curve which has been given to represent correctly the distribution of errors, the mean error of the mean of N observations can be proved to be $m/\sqrt{N}$ if m is the mean error of a single observation. This result is important, for it shews that we may by repeating the observations and taking the mean easily double or even treble the reliability of the result, but that a very large number of observations have to be taken if we aim at a materially greater improvement.

The curve which gives the distribution of errors in a series of figures which do not represent single observations, but each of which already represents the mean of a certain number of them, has the same form as that already described for the single observations. The mean error of the final mean again represents an error such that smaller and greater errors occur equally often. By finding the total area which lies on the right-hand side of such an ordinate as FK (Fig. 1) we find the probability of an error greater than OK, or in other words, we find the ratio n/N, where N is the total number and n the number of errors greater than OK. It may be proved in this way that an error equal or greater than $2m$, *i.e.* double the mean error, will happen on the average in one out of every five cases, and may therefore be said to occur frequently. An error exceeding $3m$ will only occur once in 23 cases on the average, while an error of $5m$ or more will only occur once in 1300 cases. We may say that we have a reasonable security though no absolute certainty that the result will not be affected by an error greater than four times the magnitude of the mean error, the chance of greater error being in that case less than one per cent. It follows that one must take the mean of 16 observations in order to be reasonably certain that the error of the result shall not exceed the mean error of a single observation.

Let $x_1, x_2 \dots x_n$ represent the errors, *i.e.* the difference between the individual observations and their mean value. Then if $\Sigma x^2 = x_1{}^2 + x_2{}^2 + \dots + x_n{}^2$, the "mean" m and the "probable" error r of a single observation are given by

$$m = \sqrt{\frac{\Sigma x^2}{n-1}}, \quad r = \cdot 6745 \sqrt{\frac{\Sigma x^2}{n-1}}$$

and the mean M and the probable error R of their mean by

$$M = \sqrt{\frac{\Sigma x^2}{n(n-1)}} \text{ and } R = \cdot 6745 \sqrt{\frac{\Sigma x^2}{n(n-1)}},$$

n being the number of observations.

A shorter method of dealing with the observations when a less accurate result is sufficient, is to arrange them in order of magnitude and take the middle one or the mean of the two middle ones as the result. This is known as the "median." To get the mean error find in the same way the middle observation of each series above and below the median respectively. These are known as the upper and lower "quartiles." Half the difference between the two quartiles is the probable error and one and a half times this the mean error. Consult a work on the Theory of Statistics for proof, *e.g.*, Yule's *Statistics*, pp. 306–311.

Errors of quantities not directly observed.

Sometimes the quantity to be determined by an experiment is not that which is directly measured, but is deduced by calculation from the measurement. We must then calculate the error produced in the result by a certain error in the measurement. A simple case of this kind occurs if we wish to determine the area of a circle by measuring its diameter. If the diameter is D the area A is known to be $\frac{1}{4}\pi D^2$, but if an error of observation d has been committed so that the measured diameter was found to be $D + d$, the calculated area would be $\frac{1}{4}\pi\,(D + d)^2$ or

$$A + a = \tfrac{1}{4}\pi D^2 \left(1 + 2\,\frac{d}{D} + \frac{d^2}{D^2}\right),$$

where a is the error in the measured area.

If d is small so that the square of d/D may be neglected, we find

$$\frac{a}{A} = \frac{2d}{D}.$$

If an error d has been committed in a measurement D, it is usual to call the quantity $100d/D$ the percentage error. We may similarly call d/D the "per unit error" or better, the "fractional error." The fractional error of the calculated area is therefore twice the fractional error of the measured diameter.

More generally if x be a measured quantity and $y = x^{\pm n}$, the fractional error of y will be n times the fractional error of x. Hence the importance in such determinations as the coefficient of viscosity (Section XIV) of measuring as accurately as possible the radius of the tube, the fourth power of which enters into the calculation.

As a further example we may take the measurement of an electric current by a tangent galvanometer. If the angle of deflection is θ and the current C, the theory of the instrument gives the relation

$$C = \kappa \tan \theta,$$

where κ is the constant of the instrument.

If an error δ is committed in the measurement of θ, and c be the resulting error in the calculated value of C, we should have

$$C + c = \kappa \tan (\theta + \delta)$$

$$= \kappa \frac{\tan \theta + \tan \delta}{1 - \tan \theta \tan \delta}.$$

Since $\tan \delta$ is small we may neglect its square and substitute δ for its tangent, the above expression then transforms easily into

$$C + c = \kappa \tan \theta + \kappa\delta \sec^2 \theta,$$

$$\therefore\ c = \kappa\delta \sec^2 \theta.$$

Hence

$$\frac{c}{C} = \frac{2\delta}{\sin 2\theta}.$$

The fractional error of the calculated current will therefore always be larger than twice the fractional error of the measured angle because $\sin 2\theta$ will always be less than one. The error will be smallest when $\theta = 45°$, hence there is an advantage when measuring a current by a tangent galvanometer in selecting the instrument so that the angle of deflection is nearly 45°.

More difficult still are the cases in which there are two or more unknown quantities, connected by known relations. A simple case of this kind occurs when we wish to measure the time of vibration of a magnet, for instance by observing the times of passage through the position of equilibrium. If τ is the time of first passage and we only observe passages in the same direction, the successive recurrences of the event take place at times τ, $\tau + T$, $\tau + 2T$, &c. The observed times are t_1, t_2, t_3, &c., and if the errors are x_1, x_2, x_3, &c. we have the following equations

$$\begin{aligned} t_1 - x_1 &= \tau, \\ t_2 - x_2 &= \tau + T, \\ t_3 - x_3 &= \tau + 2T, \\ &\vdots \\ t_n - x_n &= \tau + (n-1)T. \end{aligned}$$

These equations contain the two unknown quantities τ and T and the n unknown values of the errors. There are therefore two more unknown quantities than equations. Any assumed value of τ and T might be made to fit into the equations if there is no limitation to the magnitude of the error. The most probable of all possible values of τ and T are those which give that distribution of errors x which we have discussed and represented by the curve in Fig. 1. The theory of probability shews that this is equivalent to saying that the most probable values of τ and T are those which give the smallest values for Σx^2, where

$$\Sigma x^2 = x_1{}^2 + x_2{}^2 + \ldots + x_n{}^2.$$

The method of calculation which allows us to determine the unknown quantities under the condition that Σx^2 shall be a minimum is called the "Method of least squares." As there will be no occasion to reduce any of the observations described in this volume by this method, we need not enter into a fuller discussion, but students interested in the subject are referred to Merriman's *Method of Least Squares*, or Yule's *Statistics*.

Errors of observation which may be eliminated by taking a large number of measurements, are comparatively easy to deal with, but the experimentalist has also to guard against the more serious danger of being misled by "systematic errors" or errors which always affect the result in the same direction. These errors may be due to a faulty arrangement of the experiment, to defective instruments and also to a bias of judgment which causes the observer to commit errors which all tend to lie on the same side of the correct value. If observations are subject to errors of the last kind they are said to be affected by a "personal equation." For instance, if a star is observed as it traverses the field of view of a transit instrument, and an observer is asked to make some sign at the moment the star passes behind a wire fixed in the instrument, it has been ascertained that most observers will signal the passage before it has actually occurred. With practised observers the difference in time between the actual and observed transit is always nearly the same, and is called the personal equation of the observer. Personal equations of small amount probably exist in many kinds of observation, such as the reading of a Vernier. It would follow that in such cases the symmetrical curve of Fig. 1 only represents approximately the distribution of errors. But personal equations need only be taken into account in cases of extreme refinement of measurement and are only mentioned here in order to point out the existence of errors of observation which are not eliminated by the usual method of multiplying measurements.

Interpolation and Empirical Formulae.

Sometimes the value of a quantity a_x has been observed for values of $x = 0, 1, 2, 3$, &c., at equal distances apart and its value is required for a value of x between two of the points of observation. In such cases the value required may be found approximately by plotting the observations and drawing a smooth curve through them, or, more accurately, by interpolation. Thus if the observed values are a_0, a_1, a_2, a_3, &c., at $x = 0, 1, 2, 3 \ldots$ and successive differences are formed as follows where $b_0 = a_0 - a_1$, $c_0 = b_0 - b_1$, &c.

x					
0	a_0				
		b_0			
1	a_1		c_0		
		b_1		d_0	
2	a_2		c_1		e_0
		b_2		d_1	
3	a_3		c_2		
		b_3			
4	a_4				

We have
$$a_1 = a_0 + b_0,$$
$$a_2 = a_0 + 2b_0 + c_0,$$
$$a_3 = a_0 + 3b_0 + 3c_0 + d_0,$$
&c.,

and generally whether x is integral or not

$$a_x = a_0 + xb_0 + \frac{x(x-1)}{1.2} c_0 + \frac{x(x-1)(x-2)}{1.2.3} d_0 + \dots$$

$$\text{or } = a_0 + \left(b_0 - \frac{c_0}{2} + \frac{d_0}{3} - \dots\right) x + \left(\frac{c_0}{2} - \frac{d_0}{2} + \frac{11e_0}{24} \dots\right) x^2 + \dots.$$

Also
$$\frac{da_x}{dx} = b_0 - \frac{c_0}{2} + \frac{d_0}{3} - \dots + \left(c_0 - d_0 + \frac{11e_0}{12} \dots\right) x + \dots$$
$$+ \left(\frac{d_0}{2} - \frac{3e_0}{4} \dots\right) x^2 + \dots.$$

If the differences in any column vanish a_x can be represented by a polynomial in x of degree one less than the order of the zero differences.

If the successive differences shew no signs of vanishing, difference tables with $\frac{1}{a_x}$, $\log a_x$ or any other function of a_x may be drawn up to determine whether a polynomial in x will represent the new function. See Running, *Empirical Formulas*.

When the observations are not taken at equal intervals the method of "divided differences" can be used. See Whittaker and Robinson, *Calculus of Observations*, p. 20.

When the differences are too irregular the observations should be plotted and a smooth curve drawn through them, or they should be smoothed by the processes described in Whittaker and Robinson, Chap. XI.

SECTION II

MEASUREMENT OF LENGTH

THE simplest method of measuring a length consists in the direct comparison of the length to be measured with a scale which serves as a standard, and which is subdivided into intervals of say 1 mm. If the length to be measured is not exactly a multiple of a millimetre, it becomes necessary in some way to measure or otherwise to estimate the fraction of a subdivision.

The application of a "Vernier" to this purpose is explained in elementary books (see also Schuster and Lees's *Intermediate Physics*); the Vernier can however often be dispensed with and the subdivision estimated with sufficient accuracy by the eye.

The following hints as to estimation will be found of use, but the student will only obtain confidence in his judgment by constant practice. He should not attempt to estimate more closely than to a tenth of a division, though an experienced observer will under favourable circumstances estimate correctly to a fiftieth.

The eye subdivides easily an interval into two equal parts, and if the point to be measured is, *e.g.*, at C (Fig. 3), between the scale divisions marked 2 and 3, even an unpractised observer will put it down at once as 2·5. Should the point C lie as in Fig. 4 he will in his mind fix a point C' at the same distance from C as that point is from the nearest scale division, then a point C'' again at the same distance from C'. By this means he will see at once that C is less than a third but more than a fourth part of the way from the division 2 towards the division 3; hence he will read between 2·25 and 2·33 and there is no difficulty in putting it down as 2·3.

2 C 3

FIG. 3

2 C C′ C″3

FIG. 4

In Fig. 5 an observer would see if he divided the interval between 2 and 3 into two equal parts at B, and the first half again into equal parts at B', that the point C was rather nearer to 2 than to B and he would put it down as 2·2.

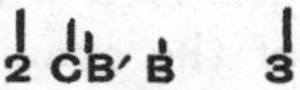

Fig. 5

The greatest liability to error occurs when the point C lies between ·1 and ·2 of an interval, and there will probably be a tendency to over-estimate the distance from the nearest division.

In order to increase the power of judgment in this case, it is advisable for the student to draw two lines A and B and a third C, so that AC is the tenth part of AB (Fig. 6). This should be done with AB varying in length between ·5 and 3 cms.

Fig. 6

If a student once has got a good idea of a subdivision into ten parts, the estimation of subdivisions in general will not present much difficulty.

Exercise. Estimate the distances of the middle lines from the left-hand lines in terms of the distance between the end lines taken as unity, in the cases shewn below. When these estimates have been entered in the note-book, measure the distances on a millimetre scale and check your estimates.

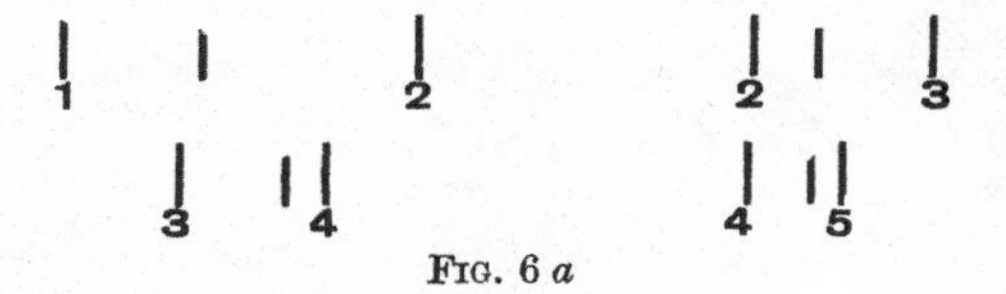

Fig. 6 *a*

When in the following exercises no special directions are given as to the method to be used in determining a length, a scale graduated in cms. and mms. should be applied to the length to be measured, adjusted to read an integral number of cms. at one end of the length, and the point on the scale corresponding to the other end, read. If this end coincides with a mm. division of the scale, the distance is read off immediately. If not, its position between two consecutive divisions must be estimated as explained above.

When the length to be measured is small, a microscope provided with a divided transparent scale in the focal plane of the eyepiece is used. If, as for instance in the viscosity exercise (Section XIV), the radius of a narrow tube is to be measured, the tube is placed vertical under the microscope and the diameter read off directly in terms of divisions of the eyepiece. The value of the scale division must be determined by placing a transparent millimetre scale under the microscope, focussing and comparing it with the eyepiece scale.

The divided scale of the eyepiece may be replaced by a "Micrometer," *i.e.* a cross wire which can be moved sideways by means of a fine screw the turns of which are read off on the screw head. Or the whole microscope may be "traversed" in a direction at right angles to its optic axis by a micrometer screw with a graduated head.

SECTION III

MEASUREMENT OF INTERVALS OF TIME

ALL time measurements in a Physical Laboratory are referred to some standard clock, watch, or chronometer, the error of which must be known in the few instances in which absolute time is required. In the great majority of cases intervals of time only are measured, and the rate only of the clock is required.

A clock or watch is generally available which goes correctly to within a minute a day, *i.e.* one minute in 1440.

Such an instrument would therefore allow us to determine intervals of time to within less than one part in a thousand, an accuracy sufficient for many purposes. When greater accuracy is aimed at, or when the error of the timepiece is too great, the rate must be determined by comparison with some better instrument or by direct observations of the sun or a star with a sextant or transit instrument.

The method to be adopted in measuring an interval of time varies with the nature of the interval, but in most cases which occur in the laboratory the interval is that between two consecutive occurrences of some periodic event, *e.g.* the passage of a pendulum through its position of equilibrium. The simplest course to pursue is then to count the number of occurrences in a given time, say one or two or more minutes, according to the accuracy required. The number of occurrences divided by the time elapsed between the first and last gives the required interval.

If the occurrences follow each other rapidly, as *e.g.* when they take place four times a second, it requires a little practice to count them. This is most easily accomplished by taking them in groups of four, counting thus: one–two–three–four, one–two–three–four, &c., stress being laid on the four in pronouncing, and a mark being made simultaneously with a pencil on a sheet of paper. Four times the number of marks made on the paper in a given time is the number of occurrences during the time.

If the periodic time exceeds ten seconds, the total time required for a given number of occurrences is great, and the accuracy of the determination of the interval may be increased without unnecessarily increasing the time occupied, by observing separately the times of a number of occurrences. In this case a watch or clock beating seconds or half-seconds is required. The observer begins a few seconds before the occurrence to count seconds in time with the clock, while he watches for the occurrence, the counting being done thus: twenty-*one*, twenty-*two*, twenty-*three*, the twenty being spoken lightly, and stress being laid on the *two*, *three*, &c., each of which should be pronounced in coincidence with the tick of the clock. By exercising a little care the twenty may be pronounced just at the half-second, and the exact time of the occurrence fixed to less than half a second. The times of consecutive occurrences numbered 1, 2, 3, 4, 5, &c. are found in this way, and written down.

To shew how the observations are reduced we take as example the determination of a time of oscillation, 12 successive passages for instance of a galvanometer needle through its position of equilibrium in the *same direction* being observed. Let the observed times be:

Event No.	Time	Event No.	Time
1	$11^h\ 23^m\ 3^s$	7	$11^h\ 24^m\ 1^s$
2	12	8	11
3	22	9	20
4	32	10	30
5	41	11	39
6	51	12	49

If we only make use of the first and last observations we should find the interval of time for 11 oscillations to have been $1^m\ 46^s$, and therefore the time of a single oscillation to have been $9^s{\cdot}64$.

The intermediate observations are not made use of in this method of calculation, and there is therefore some loss of accuracy in the final result. To discover the method of reduction which, without too much numerical labour, should give the best result, we notice in the first instance that the probable error in the measurement of a time interval will be the same, whether that interval be large or small. If e is that probable error, and if the measured interval includes n complete oscillations, the probable error of the periodic time calculated from the n intervals will be e/n.

If instead of observing the time of n oscillations we had observed the time of a single oscillation, and repeated the observation n times, the mean of the results so obtained would have given a probable error $e/\sqrt{n}$ (Section I). As e/n is smaller than $e/\sqrt{n}$ in the ratio of $\sqrt{n} : 1$, it follows that a better result is obtained if only two observations are taken, viz. one at the beginning and one at the end of n consecutive intervals, than if n separate single intervals implying $2n$ observations are taken. This shews that it is not the number only of observations which determines the accuracy of a result, but also their intelligent arrangement and reduction. Returning to the above example it is easily seen that the result may be improved by making use of the first two and last two observations. The interval between the first and the 11th is 96^s for 10 vibrations or 9·60 for one, the interval between the second and last is 97^s for 10 vibrations or 9·70 for one, the mean being 9·65.

The probable error of each of these results separately is $e/10$ and that of their mean therefore $e/10\sqrt{2}$ or nearly $e/14$, as compared with $e/11$ in case only the first and last observations are used.

More generally taking the first p and the last p observations and taking the difference between the $(n-p+1)$th and the first, that between the $(n-p+2)$th and the second, and so on until we come to the difference between the last and the pth observation, we secure p values of $n-p$ intervals, the time of oscillation calculated from each separately will have a probable error $e/(n-p)$ and the probable error of their mean will be $e/(n-p)\sqrt{p}$. The smallest probable error is obtained when $(n-p)\sqrt{p}$ is as large as possible for a given value of n, and it is easily shewn that this is the case when $p = n/3$.

Applying this to the above example it follows that it would be advisable to reject the 5th, 6th and 7th observations and to arrange the rest as follows:

Event No.	Time	Event No.	Time	Time of 8 intervals
1	$11^h\ 23^m\ 3^s$	9	$11^h\ 24^m\ 20^s$	$1^m\ 17^s$
2	12	10	30	18
3	22	11	39	17
4	32	12	49	17
			Sum	$5^m\ 9^s$
			Mean	1 17·25
			∴ Interval	9·656

The observations which are not made use of need not be written down at all, but in that case great care is needed to avoid making an error in the number of observations which are omitted. The liability to error is reduced, and the reliability of the result not materially interfered with by a slightly different arrangement. If p intervals are observed, *i.e.* $p + 1$ observations taken, then p intervals omitted, and p again observed, the time when the last set of p observations should commence may be readily calculated. Thus in the above example if four intervals had been measured the difference of the first and fifth observation would have given 38^s for the time of four intervals, adding 38^s to the fifth observation the calculated time for the ninth would have been $11^h\ 23^m\ 41^s + 38^s = 11^h\ 24^m\ 19^s$, the observer could therefore have rested till near that time and then carefully watched for the occurrence, which as the table shews actually took place at $24^m\ 20^s$.

The method which has been explained in connection with the measurement of time intervals is equally applicable in other cases, such as that of temperatures measured with a uniformly rising or falling thermometer read off at equal intervals of time, the rate of fall or rise of the thermometer being the quantity to be determined (Section XXI). Sections XVII and XVIII also shew examples of the same method of reduction.

SECTION IV

CALIBRATION OF A SPIRIT LEVEL

Apparatus required: ***Spirit Level, tilting board with levelling screw, screw gauge.***

The Spirit Level consists of a slightly bent tube (Fig. 7, in which the curvature is exaggerated) partly filled with alcohol. The bubble of air or of vapour of alcohol left in the tube will always set itself so that it is at the *highest point* of the tube. The tube is generally divided as shewn in Fig. 8, so that it is easily seen when the bubble is in the centre of the tube. If the level is in proper adjustment the base of the plate to which the level is attached should be horizontal when the bubble is in the central position. It is therefore necessary that the central point D of the tube (Fig. 7) should be farthest from the supporting plate AB. An adjusting screw T is generally provided at one end of the level, by means of which the tube can be tilted with respect to the base to secure that this shall be the case. If the table on which the level is placed is nearly but not quite horizontal, the bubble may not stand in the middle, but be slightly displaced; the position of its centre may be read off in scale divisions, and if the value of the scale division is known, the inclination of the table to the horizontal may be calculated. It is the object of the present exercise to calibrate the divisions of a level so as to find their values in angular measure, and also to shew how even if the spirit level is not itself in proper adjustment it may be used to determine whether a surface is horizontal.

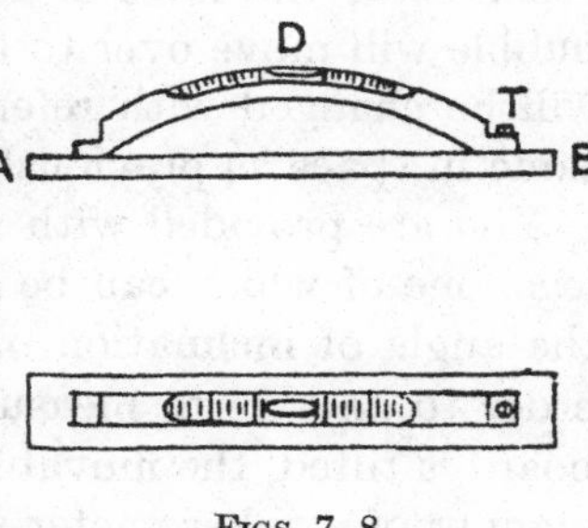

FIGS. 7, 8

If the level has not been tested and when placed on a table shews the centre of its bubble shifted to one side, it is uncertain whether this is due to the fact that the level is not in adjustment or that the table is not horizontal. But if the level be turned through 180° round a vertical axis so that the points A and B are interchanged, we may find out which is the correct interpretation. If the table is horizontal and the level wrong, the bubble should remain at the same place, for that point of the tube which is farthest from AB will remain so in whatever position the level is placed. If on the other hand the level is correct and the table inclined, the bubble will move over to the other side and its new position will be changed with reference to the level, but will be the same in space as previously.

You are provided with a tilting board supported on three legs, one of which can be screwed up or down so as to alter the angle of inclination of the board to the horizontal. In order to be able to measure the amount through which the board is tilted, the movable leg carries a divided circle in the manner of a spherometer screw. The rotation of the screw is read on the divided head and its pitch must be separately determined. The board is to be placed with its legs on thick

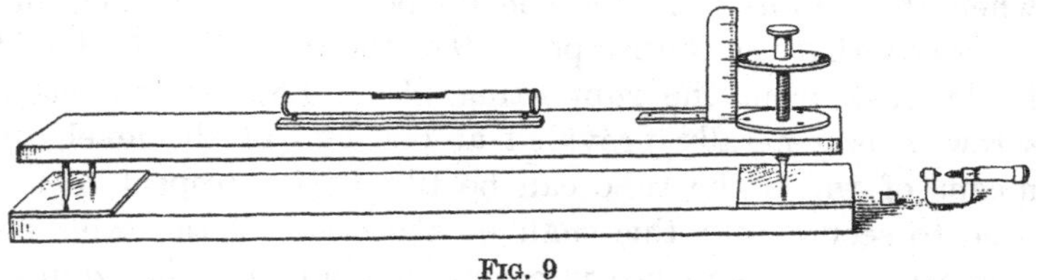

FIG. 9

glass plates with their upper surfaces parallel to each other, or better in the same plane, and the Spirit Level placed on the marked space at the centre of the board, with its length perpendicular to the line joining the fixed legs of the board. See that the board supports the level at its ends, so that the bubble is not disturbed when one end of the level is pressed down gently with the finger.

One end of the level is provided with an adjusting screw. Note the end at which this adjusting screw is placed. For the sake of uniformity begin with it at the right-hand end. If the level has no adjusting screw, mark one end in some way and place it to the right.

Level the board by means of its screw till the bubble is nearly in the centre.

Let the readings of the ends of the bubble be, *e.g.* – ·4 and + ·5, the divisions at the screw end of the level being called positive, those at the other end negative.

Now remove the level and reverse it so that the screw end is to the left. If the reading of the screw end of the bubble remains + ·5, the board is horizontal. If the reading is altered to + ·6 say, take out half the error by means of the board screw, so that the end of the bubble now reads + ·55.

Reverse the level again so that its screw is on the right hand.

This reversal should not alter the reading of the screw end of the bubble from ·55.

If it does, again take out half the error by means of the board screw, and reverse again.

In this way a position of the board screw is found, such that reversal of the level produces no change in its readings.

The board is then horizontal.

Adjust the screw of the level till, when the level is on the board, the two ends of the bubble are at equal distances from the centre of the scale.

The level is then in proper adjustment for use.

If the level provided has no adjusting screw, the error of the level is thus determined and the calibration may be proceeded with.

To determine the value in degrees, minutes, and seconds, of each division of the level, place it with its screw end to the right. Adjust the board screw till the centre of the bubble is at 0, and read the screw head.

Now, adjust the board screw till the centre of the bubble is at 1 to the right of zero, and read the screw head, recording the number of turns and decimals of a turn. Repeat, placing the centre of the bubble at 2, and so on for the other divisions of the right-hand scale.

Again adjust the centre of the bubble to 0, read the screw head, and take observations as the bubble is moved to the left.

Determine the pitch of the screw by first adjusting the screw till the centre of the bubble is at zero, reading the screw head, then inserting under the screw the small piece of plate-glass about 5 mms. thick provided, again adjusting the screw till the bubble is at zero, then reading the head and recording the number of turns and parts necessary to make the adjustment. Determine by the screw gauge the thickness of the glass-plate, and thence calculate the pitch of the board screw.

Measure the perpendicular distance l from the screw leg to the line joining the other two legs of the board.

Then if n = number of turns of board screw from its reading when the bubble was at 0, and p = pitch of the screw,

np = distance through which screw has been raised,

and np/l = angle in circular measure through which the board has been tilted (since the angle is small),

or in angular measure

$$\theta = \frac{180^\circ}{\pi}\frac{np}{l}.$$

Determine in this way the angle of tilt for each observation, and draw up a table as follows:

Tilting board A.

Distance of screw from line joining fixed legs = 24·16 cms.
Reading of screw head when glass inserted = 10·515 turns.
" " " no glass = ·405 turn.
Thickness of glass = ·507 cm.
Pitch p of screw of board ·507/10·11 = ·0502 cm.

$$\frac{180^\circ}{3\cdot14}\cdot\frac{p}{l} = \frac{180 \times \cdot0502}{3\cdot14 \times 24\cdot2} = \cdot121.$$

Spirit Level A.

Readings		n	$\frac{180p}{3{\cdot}14l}\cdot n$	Inclination to horizontal	
centre of bubble	screw head			θ	$\delta\theta$
0 screw end	·100	0	0	0	—
1 ,,	1·388	1·288	·156	9′ 22″	9′ 22″
2 ,,	2·640	2·540	·307	18′ 25″	9′ 3″
3 ,,	3·763	3·663	·443	26′ 36″	8′ 11″
4 ,,	4·537	4·437	·536	32′ 11″	5′ 45″
5 ,,	5·163	5·063	·612	36′ 42″	4′ 31″
6 ,,	5·683	5·583	·676	40′ 34″	3′ 52″

Similarly for readings of the bubble on the other end of the scale.

The column headed $\delta\theta$ gives the angle through which the level has to be tilted in order to shift the bubble through one division. If the curvature of the tube of the level is constant, so that its axis is an arc of a circle, the numbers in the last column should be the same. The particular level to which the above numbers refer had a smaller curvature nearer the end than at the centre.

Students may find that the length of the bubble is not constant owing to the evaporation or condensation of the alcohol due to changes of temperature. Hence the necessity for reading both ends of the bubble.

Tilting tables of metal are now made, in which the pitch of the screw and distance of the screw from the fixed legs are so chosen that the screw head can be graduated directly in minutes and seconds of arc.

SECTION V

CALIBRATION OF A GRADUATED TUBE

Apparatus required: *Graduated tube, mirror, lens to be used as magnifying-glass, and clean mercury.*

The graduations on a tube which is to be used for measurements are as a rule placed at equal distances apart along the tube, but if the tube is used to measure volumes, the assumption that distances between the graduations represent equal volumes of the tube will lead to error unless the tube is of uniform bore. It is the object of calibration to find the correction which must be applied to the reading on the scale of equal lengths, in order to convert it into one of equal volumes. The calibration correction is that part of a division which must be *added* to any reading on the equal length scale in order to obtain the reading of the same point on an equal volume scale.

Let a tube be provided with equidistant graduations, 1, 2, 3, &c., see Fig. 10, and imagine graduations I., II., III., &c., along the same tube, such that the intervals between each represent not equal lengths, but equal volumes of the bore of the tube.

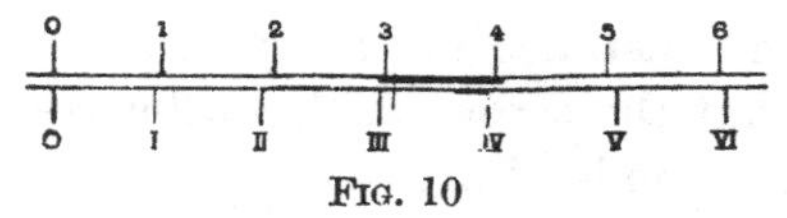

FIG. 10

Thus the zeros of the scales being supposed to coincide, the division 1 on the scale of equal lengths corresponds to rather more than I. on the equal volume scale line, the correction at the point 1 is therefore positive. Similarly the corrections at points 2, 3 and 4 are positive, and at 5 and 6 negative.

Let the corrections at successive divisions be x_0, x_1, x_2, x_3, &c., and suppose a mercury thread, which if the temperature is constant will be of constant volume, say about equal to that of a scale division, to be pushed along the tube.

In Fig. 10 the ends of the thread lie approximately at the points marked 3 and 4. Let δ_4^3 be the length of the thread when placed between the points 3 and 4. The point 3 of the equal length scale would read $3 + x_3$ on the equal volume scale, the point 4 would read $4 + x_4$, the true volume between these points would be therefore, not 1 but $1 + x_4 - x_3$. Similarly if a were the number of divisions between points marked $3a$ and $4a$ on an equal length scale, the subdivisions being left out in Fig. 10, the true length of the interval would be not a but $a + x_4 - x_3$.

The excess of δ_4^3 over a read on the equal length scale being small, will not be sensibly different when read on the equal volume scale, hence the corrected length of the thread becomes $x_4 - x_3 + \delta_4^3$ and this we know must be equal to the constant volume of the thread, say l, wherever the thread is placed.

Applying this equation successively to the different positions of the thread we obtain the following n equations,

$$\left.\begin{array}{l} x_1 - x_0 + \delta_1^0 = l \\ x_2 - x_1 + \delta_2^1 = l \\ \vdots \\ x_n - x_{n-1} + \delta_n^{n-1} = l \end{array}\right\} \quad \ldots\ldots\ldots\ldots\ldots\ldots(1)$$

involving $n + 2$ unknown quantities, $x_0, x_1, \ldots x_n, l$. Hence we must know two of the corrections, or at any rate two relations between them, to solve the equations.

An important case of calibration is that of a thermometer between freezing point and boiling point, the corrections at these two points being determined by experiment, *i.e.*, x_0 and x_n are known, and there only remain n unknown quantities.

In that case, to solve the above set of equations, add them up and we get

$$x_n - x_0 + \text{sum of all } \delta\text{s} = nl \quad \ldots\ldots\ldots\ldots(2).$$

$$\therefore l = \frac{x_n - x_0 + \text{sum of } \delta\text{s}}{n}$$

$$= \delta, \text{ say.}$$

Hence from (1)

$$\left.\begin{aligned} x_1 &= x_0 + \delta - \delta_1{}^0, \\ x_2 &= x_1 + \delta - \delta_2{}^1 = x_0 + (\delta - \delta_1{}^0) + (\delta - \delta_2{}^1), \\ x_3 &= x_2 + \delta - \delta_3{}^2 = x_0 + (\delta - \delta_1{}^0) + (\delta - \delta_2{}^1) + (\delta - \delta_3{}^2), \\ &\vdots\ \cdots\cdots\cdots\cdots\cdots\cdots\cdots\cdots\cdots\cdots\cdots\cdots \\ x_n &= \qquad\qquad = x_0 + (\delta - \delta_1{}^0) + \ldots\ \ldots + (\delta - \delta_n{}^{n-1}) \end{aligned}\right\} (3).$$

From which, since x_n and x_0 are given, x_1 may be calculated from the first equation since $\delta_1{}^0$ has been determined. Having found x_1, x_2 is calculated from the next equation, and so on successively for all the corrections. If the calculations have been carried out correctly the calculated value of x_n should agree with that assumed.

A tube is given to you divided into millimetres (Fig. 11).

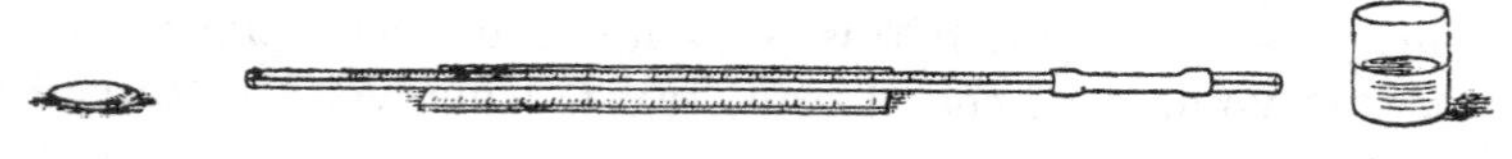

Fig. 11

It is to be calibrated between points 16 cms. apart, the errors being required for points at a distance of 2 cms. apart. That is to say, assuming the first and seventeenth centimetre division to be known, the errors at the third, fifth, &c. division are to be found.

The most difficult part of the operation consists in obtaining a mercury thread of the required length. It may most easily be accomplished by slipping a short piece of india-rubber tubing over one end of the glass tube. The free end of the rubber tube is then compressed between the finger and thumb of the right hand, so as to close it completely, the end of the glass capillary tube dipped into clean mercury, and then the rest of the rubber tube compressed between the finger and thumb of the left hand to expel a little air from it. On releasing the left hand, a thread of mercury is drawn up.

The capillary tube should then be quickly placed nearly horizontal with the lower end over the mercury in the bottle, and the end of the rubber tube released. By gently tilting the tube, the thread may be moved from end to end of the tube, and it should be noted whether it moves freely without leaving portions of mercury behind, or the ends of the thread ceasing to be convex. If this is not the case the thread should be removed from the tube and the tube cleaned, first with a little dilute acid, then with water, then with alcohol, and thoroughly dried. A thread of mercury should then be drawn up into the tube, made to approach the lower end of the tube, and a small drop of mercury forced out and cut off with the finger-nail or a knife.

By carrying out this operation several times the thread may be diminished in length till it occupies between 19 and 21 of the small scale divisions, *i.e.* within 1 mm. of 2 centimetres.

When this is the case, move the thread till its left-hand end nearly coincides with the division on the left at which the calibration is to commence.

Place the tube on a strip of mirror glass and read both ends of the mercury thread by means of a magnifying lens, avoiding parallax by placing the eye so that its image is covered by the part read or so that the graduation nearest the end of the thread covers its image in the mercury, and then slipping the lens into position between the eye and the scale. The excess of the observed length of the thread over 20 scale divisions is the $\delta_1{}^0$ of the preceding equations.

Now move the thread forward 20 small scale divisions so that its left-hand end nearly occupies the position of the right-hand end in the previous case, and obtain $\delta_2{}^1$ as before.

Continue the operation till the ends have been read in eight successive positions, and then take readings as the thread is moved backwards. The means of the δs observed at each part of the scale going and returning should be used in the calculations, which should be carried out and tabulated as shewn below.

A check for some of the intermediate points may be obtained by taking threads 4 and 8 cms. long. Calling the first division 0, the 4 cms. thread will give the corrections at the points 4, 8 and 12, while the 8 cms. thread will only give the correction at the division 8.

To calculate the corrections it is necessary first to find δ, from x_0, x_n and the quantities $\delta_1{}^0$, $\delta_2{}^1$, &c. If these are tabulated as shewn below and δ found, the differences $\delta - \delta_1{}^0$, $\delta - \delta_2{}^1$ may be written down as in the fifth column and the corrections obtained by successive addition.

Arrange your observations and calculations as follows, giving the mark on the label attached to the tube so that it may be identified:

TUBE B. Date:

$x_0 = \cdot010$ $x_{16} = \cdot018$*

Going	Returning	Mean	$\delta - \delta_2^1$ &c.	Corrections
				$+\cdot010 = x_0$
$-\cdot04$ to $2\cdot03 = 2\cdot07$	$-\cdot08$ to $2\cdot0 = 2\cdot08$	2·075	−·0315	$-\cdot021 = x_2$
1·98 to 4·04 = 2·06		2·065	−·0215	$-\cdot043 = x_4$
3·96 to 6·02 = 2·06		2·062	−·0185	$-\cdot062 = x_6$
5·98 to 8·04 = 2·06		2·067	−·0235	$-\cdot085 = x_8$
8·00 to 10·03 = 2·03	&c.	2·032	+·0115	$-\cdot074 = x_{10}$
9·99 to 12·01 = 2·02		2·017	+·0265	$-\cdot047 = x_{12}$
12·01 to 14·02 = 2·01		2·017	+·0265	$-\cdot020 = x_{14}$
13·99 to 16·00 = 2·01		2·005	+·0385	$+\cdot018 = x_{16}$
	$x_{16} - x_0 =$	·008		
	Sum =	16·348		
	$\frac{\text{Sum}}{n} = \delta =$	2·0435		
				$+\cdot010 = x_0$
$-\cdot05$ to $4\cdot10 = 4\cdot15$	$-\cdot06$ to $4\cdot10 = 4\cdot16$	4·155	−·049	$-\cdot039 = x_4$
4·01 to 8·17 = 4·16		4·160	−·054	$-\cdot093 = x_8$
7·97 to 12·06 = 4·09	&c.	4·095	+·011	$-\cdot082 = x_{12}$
12·01 to 16·04 = 4·03		4·005	+·100	$+\cdot018 = x_{16}$
	$x_{16} - x_0 =$	·008		
	Sum =	16·423		
	$\frac{\text{Sum}}{n} = \delta =$	4·1057		
				$+\cdot01 = x_0$
$-\cdot03$ to $8\cdot12 = 8\cdot15$	$-\cdot04$ to $8\cdot12 = 8\cdot16$	8·155	−·096	$-\cdot086 = x_8$
7·92 to 15·89 = 7·97		7·955	+·104	$+\cdot018 = x_{16}$
	$x_{16} - x_0 =$	·008		
	Sum =	16·118		
	$\frac{\text{Sum}}{n} = \delta =$	8·059		

* These numbers are assumed given for the sake of completing the exercise.

Taking the readings along the tube as abscissae and the corrections at these readings as found by the first measurements as ordinates, above or below according as the correction is positive or negative, plot a "Calibration Curve" for the tube as shewn below, Fig. 12.

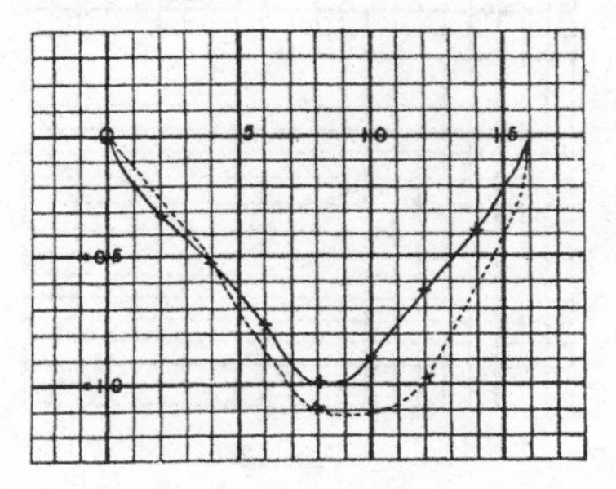

Fig. 12

Do the same for the second and third sets of readings, taking the same abscissae.

If l is the length of the mercury thread in any position, and A the mean area of the cross section of the tube within the part occupied by the mercury thread, Al is the volume of the thread. As this volume remains constant as the thread is moved, the length of the thread varies inversely as the mean area of cross section within its length.

Hence by taking again distances along the tube as abscissae, and erecting at the different points occupied by the centre of the mercury thread ordinates inversely proportional to the observed lengths of the threads, we get a representation of the way in which the cross section of the tube varies.

Centre of thread	Length of thread	Reciprocal of length
1 cms.	2·075 cms.	·482
3	·065	·484
5	·062	·485
7	·067	·484
9	·032	·492
11	·017	·495
13	·017	·496
15	2·005	·499

Plot these reciprocals as shewn Fig. 13.

The most important practical application of the preceding exercise occurs in the calibration of thermometer tubes. All thermometers used for accurate work should be calibrated

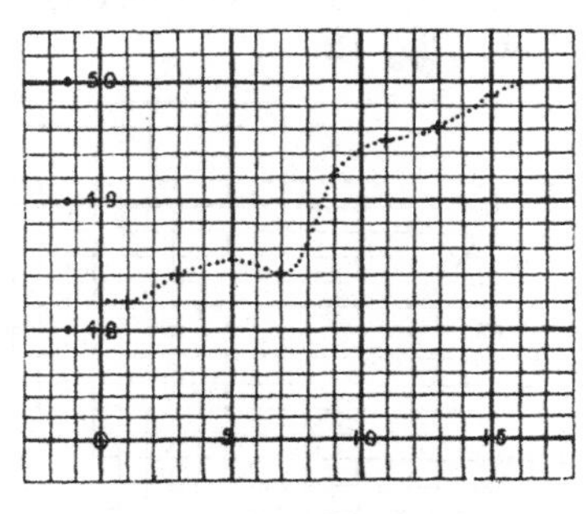

FIG. 13

either by the maker or by the observer. Information as to the methods of breaking off a mercury thread and complete methods of calibration will be found in Guillaume's *Thermométrie*.

BOOK II

GENERAL PHYSICS

SECTION VI

THE BALANCE

Apparatus required: *Delicate balance, centigram rider, two* 500 *gram weights.*

The Balance, in its simplest form, consists of a straight beam AB (Fig. 14) provided at its centre with a knife edge C on which it is supported, and carrying at its ends the pans P and Q, on which the masses to be compared are placed. If the two halves of the beam are alike in every respect, so that the centre of gravity of the beam is at the point of support C, and if the pans have equal mass, the balance will be in neutral equilibrium, whether unloaded or loaded with equal masses. If *unequal* masses are placed in the pans, the equilibrium will be unstable, however small the inequality may be.

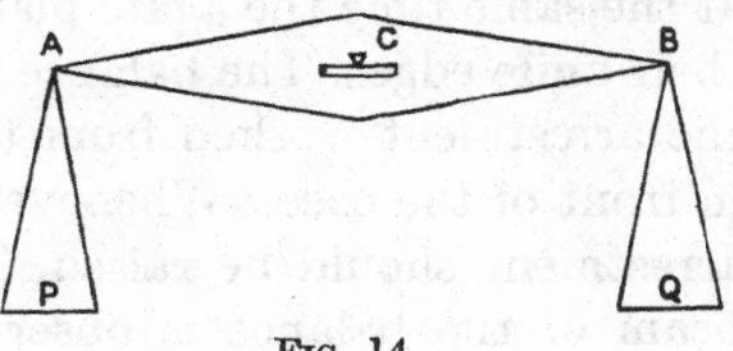

Fig. 14

It would be inconvenient to have a delicate balance constructed according to this principle, For as two masses are never exactly equal, the balance would never be in equilibrium. A delicate balance, to be useful, should allow us to determine the difference between two nearly equal masses, and it is proved in treatises on Mechanics that this can be done by constructing the beam AB of the balance so that its centre of gravity is slightly

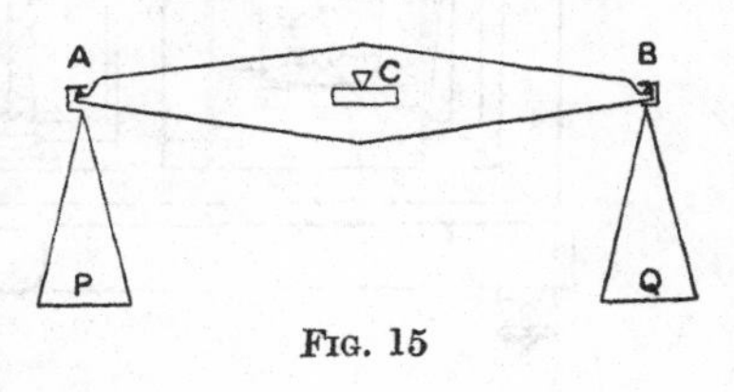

Fig. 15

below the fulcrum C. The balance will then have a stable position of rest even if the two weights P and Q are not quite equal, and the amount of inequality may be determined by an observation of this position.

The pans of a delicate balance are suspended from agate planes supported on knife edges at the ends of the beam, which is itself provided at its centre with a knife edge resting on an agate plane at the top of the pillar (Fig. 15). The accuracy which can be attained in weighing depends to a great extent on the freedom with which these three knife edges can turn on their planes. In order to save these delicate parts as much as possible from wear and tear, a so-called arrestment is provided, by means of which the central knife edge of the beam may be raised from its agate plane, while at the same time the agate planes of the pans are raised from their knife edges. The balance is enclosed in a glass case, and the arrestment worked from the outside by the screw head in front of the case. Whenever the balance is not in use the arrestment should be raised. The change of position of the beam of the balance is observed by the help of a pointer attached to the beam, the free end of which moves in front of a scale at the foot of the pillar.

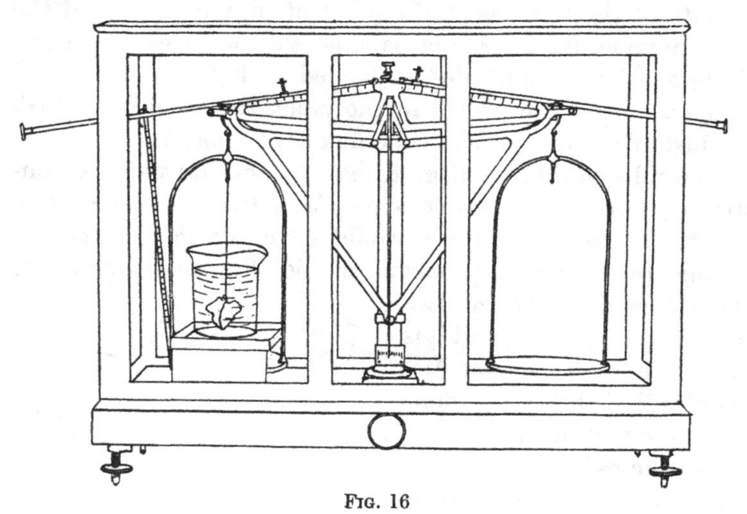

Fig. 16

Before beginning to work, students should carefully inspect the balance (without opening the case), and make in their note-book a rough sketch of the mechanism of suspension and arrestment. The reference books in the laboratory library contain detailed descriptions of the balance, which may be consulted with advantage.

Exercise I. To find the zero of the unloaded balance.

Carefully lower the arrestment of the balance, and watch the movement of the pointer. If it moves slowly through a range of from 3 to 6 divisions of the scale, observations may be commenced. If the extent of the swing is greater, it is a sign that the arrestment was not handled with sufficient care; it should be gently raised again, and lowered more slowly. If the swing is too small, open one of the windows of the balance and produce a weak current of air by gently waving your hand a few times inside the case, being careful not to bring it into contact with any part of the balance. A little practice will enable you to obtain a workable swing. The window is then shut and the observations commenced.

As it would take too long to wait until the balance has come to rest, the position of rest must be determined by observations made during its motion.

For this purpose the turning points of the pointer on the scale must be observed. If the balance moved without any friction, the pointer would move to equal distances on both sides of the position of rest, which could therefore easily be deduced from two successive readings of the turning points. Owing to friction the oscillations gradually diminish, and in order to find the position of rest more than two readings are required.

The following example will shew how the position of rest of the balance is calculated from three observations of the turning points of the pointer.

0 5 10 15 20

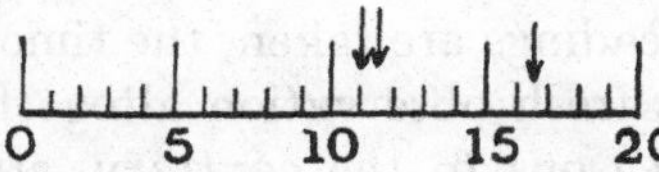

Fig. 17

Suppose that after the arrestment has been carefully lowered, the pointer moves to the right, then turns back again to the left at a point which is not observed. Let the three following turning points, indicated by arrows in Fig. 17, be read off as follows: 11·0, 16·6, 11·6, the centre of the scale over which the pointer moves being marked 10 in order to avoid negative numbers. If there were no friction the arithmetical mean between 11·0 and 16·6 would give the position of equilibrium. In reality this mean (13·8) gives too *small* a value, for without friction the needle would have moved beyond 16·6. For the same reason the mean of 16·6 and 11·6 (14·10) would give too *large* a value, and as the frictional retardation has been practically the same during the forward and during the backward swing, the errors of the two means will be equal and in opposite directions, and hence the arithmetical mean between 13·8 and 14·10, *i.e.* 13·95, gives a sufficiently correct result for the required position of rest. The same numerical result is obtained by taking the mean of the two successive swings on *one* side (11·30 in the above example) and then the mean between this number and the intermediate turning point on the *other* side. Thus the mean of 11·30 and 16·6, or 13·95, would be the true position of rest. If the numbers in the first instance be read off correctly to a tenth of a division, the final result involving three readings may be given to another decimal place, for though the observations may not be sufficiently accurate to fix the final position within a *hundredth* of a scale division, they should be sufficiently accurate to determine it more nearly than a *tenth* of a division. Practice will enable the student to read with certainty to less than a tenth of a division.

The effect of small errors of observation on the final result will be diminished by taking an uneven number of readings more than three, beginning and ending on the same side. The same method of reduction may then be employed. There is, however, a considerable loss of time when too many readings are taken, the time being better employed in taking a fresh observation altogether, and unless there are special reasons to the contrary, students in the laboratory should take only five successive readings and put them down in their note-book as in the following example, noting the temperature indicated by the thermometer in the balance case.

Balance: *A*. Date:

	Turning points		
	left	right	
	7·4 7·6 7·7	13·1 13·0	
Sums ...	22·7	26·1	
Means ...	7·57	13·05	
Position of rest	10·31		Temperature 18°·1 C.

Whenever a weighing is taken the observations should be recorded as above, but in writing out results the final positions of rest and the temperatures need only be given.

To bring the balance to rest raise the arrestment gently so that it lifts the beam when the pointer is passing through the point of rest, and thus injures the knife edge as little as possible.

Take three sets of observations of the position of rest of the balance, raising the arrestment between each set. Find the mean value of the zero as determined by the three sets.

The temperature of the balance case is required in this and the next exercise, in order that it may be ascertained by observations taken on different days whether the zero of the balance shews any appreciable change depending on temperature. For ordinary purposes it is not usual or necessary to observe the temperature.

Exercise II. To find the sensibility of the unloaded balance.

The change produced in the position of rest of an unloaded balance by 1 milligram excess of weight on one side, is known as the "sensibility" of the unloaded balance. The sensibility varies to some extent with the load placed on the pans. In order to produce a small excess of weight (less than 1 centigram) a *rider* is provided, which, by means of a sliding rod, may be shifted along the beam. The weight of the rider is 1 centigram*, but the nearer it is to the fulcrum of the balance, the smaller will be its turning moment; each half of the beam is divided into 10 equal parts marked 1, 2... from the centre to the end, so that if the rider stands at say 2, its effect is the same as the addition of 2 milligrams to the weights on the pan.

To find the sensibility it will in general be sufficient, by means of the rider, to add to one side a weight equivalent to 1 mgrm. (or for sake of greater accuracy 2 mgrs. dividing the resulting change of position of the pointer by two). In order to obtain sufficient practice in accurate weighing, the student is required to determine the positions of rest, varying the excess of weight from 8 mgrs. on one side, to 8 mgrs. on the other of the centre of the beam. If, however, the balance is very sensitive, it may not be possible to keep the swings within the limit prescribed above (6 scale divisions), when an excess of 6 or 8 mgrs. is placed on one side.

The observations should be taken and recorded as in Exercise I. The numbers obtained should be summarised in the Book of Results as follows:

* The weights of riders sold commercially often differ so much from their nominal value that serious errors may be introduced if they are used without being tested.

Balance: A. Date:

Time, 10 h. 10 m. Temperature of Balance Case, 18°·2 C.				Sensibility
Mgs. excess of Weight		Position of Rest	Difference	
	0	10·07		
On left side:	8	12·79		
	6	12·17	·62	
	4	11·43	·74	
	2	10·81	·62	
	0	10·13	·68 }	·338
On right side:	2	9·46	·67 }	
	4	8·81	·65	
	6	8·11	·70	
	8	7·42	·69	
	0	10·15		
Temperature of Balance Case, 18°·9 C. Time, 10 h. 50 m.				

Position of rest at beginning of experiment ... 10·07
" " end " ... ... 10·15
Temperature at beginning " ... ... 18°·2 C.
" " end " ... ... 18°·9
Sensibility of balance ... ... ·338

Note.—In calculating the sensibility of the balance, the only numbers in the above table taken into account, are those which give the deflections produced by 2 mgrs. on either side of the zero of the balance.

The sensibility of the balance may be increased or diminished by raising or lowering the centre of gravity of the beam, and each balance is provided with an adjustment for this purpose. But it must be remembered that an increase in the sensibility does not necessarily mean an increase in the accuracy with which a weighing can be made. As the balance is made more sensitive, its time of vibration increases, each weighing consequently takes a longer time, and small changes in the zero of the balance may take place owing to changes of temperature or other causes. The longer the time, the greater the probability of the occurrence of such disturbances. Moreover, the extra time thus spent might just as well be spent in repeating the weighing with the former sensibility, and the result would probably be improved more in this way than in the other. There is always a limit to the accuracy with which a balance will weigh, and once that degree of accuracy has been attained by a careful reading of the position of rest, it does more harm than good to attempt to increase the sensibility. Students will obtain the best results by carefully practising the method of obtaining the position of rest which has been explained above, depending on accuracy of reading rather than on a great sensibility of their balance.

Exercise III. To find the zero of the loaded balance.

If two exactly equal masses were available, to find the zero of the balance when loaded with them, it would simply be necessary to find the position of rest when the masses were placed in the pans. As however absolute equality is not easily attainable, we must find a way to determine the zero notwithstanding the inequality of the masses. If the position of rest has been found with two weights, P and Q, in the pans, then if the weights were equal, on interchanging them the position of rest would be exactly the same, and this would be the case independently of any adjustment of the balance. Even if the arms are not equally long, or if they alter their length by bending, two equal weights may still be interchanged without change in the position of rest.

But suppose the position of rest is 12·3 when P is on the left-hand side, and 10·7 when P is on the right. If Q were slightly increased, the first number would be diminished and the second number increased by *equal* amounts, and the difference in the two readings would diminish. It follows that if the increase in Q were such that the new position of rest is the arithmetical mean between 12·3 and 10·7, *i.e.* 11·5, no change would be produced on interchanging the weights, and hence 11·5 is the position which the balance would take up if equal weights P were placed in the pans.

If the sensibility of the balance were known, the experiment would give the difference between P and Q, which is evidently the weight required to be added to Q in order to produce a difference of $12{\cdot}3 - 11{\cdot}5 = {\cdot}8$ scale division.

Two masses weighing nearly 500 grams are provided. Compare the position of rest when the pans are empty, with the position of rest when these equal masses are placed in them.

Observe the temperature.

Take one observation without weights, then one with the weight marked 1 in the left and that marked 2 in the right pan, one with the weights reversed, then one with the weights as at first, then one without weights.

Again observe the temperature.

Note.—If the masses are not placed quite centrally on the scale pans, the pans will oscillate, and errors may thus be introduced. In accurate weighing this should be avoided as much as possible, by arranging the weights symmetrically. Oscillations may be stopped before the arrestment is lowered, by carefully placing the open hand so as just to touch the pan with thumb and forefinger, or by touching it with a camel hair brush.

The results are entered as follows:

Balance: A. Date:

Temperature	18°·5 C.
Zero of balance without load	11·34
Weight no. 1 on left pan; position of rest ...	10·10
,, 2 ,, ,, ...	11·42
,, 1 ,, ,, ...	10·14
Zero of balance without load	11·44
Temperature	18°·9 C.
Mean temperature	18°·7 C.
Zero of balance for load 500 grs....	10·77*
Mean zero of balance without load	11·39
Change of zero	·62
Change of temperature	·4° C.

* This number is obtained by combining the mean of the readings found with weight no. 1 on the left pan, with the reading found when that weight was on the right pan.

Exercise IV. To find the sensibility of the loaded balance.

In Fig. 15 the three knife edges are represented as being in the same straight line, but this is not necessarily the case. Let A and B, Fig. 18, represent the knife edges from which the pans are suspended, C the knife edge on which the beam of the balance rests, and G the centre of gravity of the beam alone. Draw through A and B lines perpendicular to CG that these cut CG at points k and k' below C. Let the distance of G from C be h. Then if P be the weight of the pan and contents, and a the length of the arm on the left, and Q that of pan and contents, and b the arm on the right, and if θ be the angle between CG and the vertical, we have, taking moments about C,

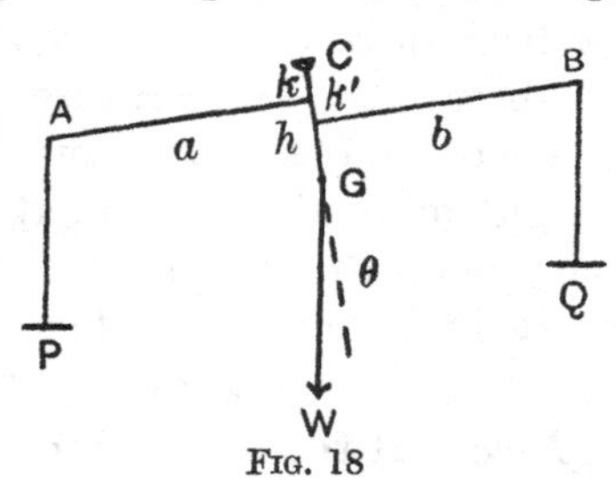

FIG. 18

$$P\,(a \cos \theta - k \sin \theta) = Q\,(b \cos \theta + k' \sin \theta) + Wh \sin \theta,$$

or

$$\tan \theta = \frac{Pa - Qb}{Pk + Qk' + Wh}.$$

If by means of the moveable vane attached to the beam the centre of gravity G of the beam is adjusted so that CG is perpendicular to AB, $k = k'$ and we have

$$\tan \theta = \frac{Pa - Qb}{(P + Q)\,k + Wh}.$$

If $Q = P + p$, then since as a rule b will be equal to a, and k will be small, the expression may be written, neglecting pk in the denominator:

$$\tan \theta = \frac{pa}{2Pk + Wh}.$$

The deflection θ produced by a given excess of weight p on one side will therefore diminish with the load if k is positive, and increase with the load if k is negative. But the value of k itself will vary with the load, as the beam of the balance always bends a little, and the pan knife edges descend as the load is increased. The makers sometimes adjust the knife edges so that the line AB is slightly above C when the balance is unloaded, and below C when the balance is loaded with more than half the maximum load for which it is intended. In that case the sensibility will first increase and will then decrease with the load.

A nut moving on a vertical screw above the beam is generally provided by the makers so as to allow a small adjustment of the position of G.

Determine the positions of rest of the balance under the following conditions:

(1) The two 500 gram weights in the pans.
(2) An excess of 2 mgrs. on the left. (3) As in (1).
(4) An excess of 2 mgrs. on the right. (5) As in (1).

Enter your results as follows:

Balance: A. Date:

Time at beginning of experiment, 11 h. 15 m.

Position of rest

(1)	with load of 500 grms.	7·98		
(2)	,, + excess of 2 mgrs. on left	8·38	8·04 (1)–(3)	·34
(3)	,, ...	8·10		
(4)	,, + excess of 2 mgrs. on right	7·84	8·19 (3)–(5)	·35
(5)	,, ...	8·28		

Time at end of experiment, 11 h. 32 m.
Mean sensibility ·172 scale division per mgr.

Exercise V. To determine the ratio of the arms of the balance.

If we assume that the ratio of the lengths of the arms of a balance is not affected by the bending of the beams, the change of zero with the load can only be due to an inequality in the arms of the balance, and the data obtained in Exercises III. and IV. are sufficient to determine that ratio. In Exercise III. the change of zero from no load to one of 500 grams was found to be ·62 division, and in Exercise IV.

the sensibility was found to be ·17; hence to bring the loaded balance with its zero 10·77 to the unloaded zero 11·39 we should have to add $\frac{\cdot 62}{\cdot 17} = 3\cdot 6$ mgrs. on the left-hand side. Let this quantity be denoted by p, and let the lengths of the left and right arms of the balance be a and b respectively. When the loaded balance comes to rest, the moments which act on the beam will be the same as in the case of the unloaded balance, with the addition of $(P + p)\,a$ on the left side, and Pb on the right side, and if the position of equilibrium is unaltered it follows that

$$(P + p)\,a = Pb,$$

or

$$\frac{b}{a} = 1 + \frac{p}{P} = 1 + \frac{\cdot 0036}{500} = 1\cdot 000,0072.$$

As the beam of the balance has a length of about 22 cms., the error of adjustment of the knife edges only amounts to ·00016 cm.

SECTION VII

ACCURATE WEIGHING WITH THE BALANCE

Apparatus required: *Balance, piece of quartz, box of weights.*

The method to be adopted in weighing depends on the object for which it is carried out. Extreme accuracy always means the spending of a good deal of time on the observations, which would be wasted if from the nature of the case such accuracy were unnecessary.

In chemical analysis relative weights only are required, and an inequality of the arms of the balance will not affect these so long as the weighings are carried out on the same side, and the method of Exercise I. p. 44 may therefore be adopted. Moreover, in many cases the errors introduced by impurities in the substances or by other causes, may amount to several mgrms., and as it would then be absurd to conduct the weighing correctly to the 10th part of a mgrm., the method may be further shortened, by reading three turning points instead of five, by assuming the zero to remain constant during the series of weighings so that it need only be taken once, and especially by using a less sensitive balance having a shorter time of vibration.

If the same balance is always used, much time may be saved by determining its sensitiveness for different loads once for all. Unless the adjustment is altered, the sensitiveness will remain the same for a considerable period.

The custom of assuming the zero always to be at the centre of the scale is not to be commended, as it may cause serious errors. The student should first learn to weigh accurately irrespective of the time it takes; he will gradually learn to weigh quickly, and to know how to save time when great accuracy is not required. In all cases where absolute and not merely relative weights are required, some method should be used which eliminates all errors of the balance, and corrections must also be made for the upward pressure of the air in which the weighing is conducted. We proceed to explain the way in which this is done.

Corrections to be applied to weighings for the buoyancy of the air.

When a body is surrounded by air, it is acted on by an upward force equal to the weight of the air it displaces. Two bodies having equal masses but different densities, will occupy different volumes, and if these bodies are placed on the two pans of the balance, it will show an apparent inequality in the weights owing to the difference in the upward pressure of the air on the two bodies. On the other hand, if two masses of different densities balance each other completely when placed on the pans, they will not in reality be equal.

If M is the mass of the substance weighed, and ρ its density, its volume is M/ρ, and the upward force of the air will be equal to the weight of a mass $M\lambda/\rho$ of air, where λ is the density of the air.

Hence the resultant downward force which acts on the beam of the balance, is the same as if the surrounding air had been removed and a mass $M(1 - \lambda/\rho)$ placed in the pan. Similarly a mass W of density σ in the second pan, produces the same downward force as a mass $W(1 - \lambda/\sigma)$ suspended *in vacuo*. If the balance, supposed to possess equal arms, is in equilibrium, we have:

$$M(1 - \lambda/\rho) = W(1 - \lambda/\sigma).$$

Hence

$$M = W(1 - \lambda/\sigma)/(1 - \lambda/\rho) = W(1 - \lambda/\sigma + \lambda/\rho) \text{ approx.}$$

The last approximate result is obtained by neglecting the squares of the small quantities λ/ρ and λ/σ, which is in this case allowable (see *Intermediate Practical Physics*, pp. 14 and 15). Hence the quantity which has to be added to the apparent mass W of the weights to obtain the true mass M of the body weighed is

$$W\lambda\left(\frac{1}{\rho} - \frac{1}{\sigma}\right),$$

when σ represents the density of the weights, generally brass, and may be taken to be 8·4. The density λ will vary with the pressure and temperature of the air, and the amount of moisture present. It will be seen that the correction is positive or negative, according as the density of the substance weighed is less than or greater than that of brass. If the masses on one or both sides of the balance consist of different materials, the correction must be determined separately for each. Thus if the weights are partly of brass and partly of platinum, it may be necessary to take this into account, and the correction becomes

$$\lambda\left[\frac{W}{\rho}-\frac{W_1}{\sigma_1}-\frac{W_2}{\sigma_2}\right],$$

when W_1, σ_1 are the mass and density of the brass weights, and W_2, σ_2 those of the platinum weights.

In the example given to illustrate Exercise II. of the present section, a mass of quartz weighing nearly 200 grms. is weighed to a tenth of a milligram, *i.e.*, to about one part in two millions. The density of quartz being 2·65, the correction for buoyancy amounts to about 68 mgrms., and has therefore to be determined with an accuracy of one part in 680. This means that the temperature of the air in which the quartz is weighed must be known to $\frac{1}{3}$ of a degree, and its pressure to 1 millimetre of mercury. The amount of moisture present in the air should also be known, but no appreciable error will be committed if the air is assumed to be half saturated. It may also be verified that an error of one part in a thousand in the assumed densities of the quartz and the brass weights, would cause errors of ·1 and ·03 mgrm. respectively in the weighing. This shews how very difficult it is to weigh accurately to one part in a million.

Before passing on to the Exercises in weighing, students should read carefully through the following instructions, which must be rigidly adhered to, as otherwise the balance may be seriously damaged.

Precautions necessary in weighing with a delicate balance.

1. Test whether the balance is in working condition by lowering the arrestment carefully; the pointer should slowly swing through a few divisions only.

2. See that the box of weights is complete, then place the riders on their supporting arms.

3. Do not touch the weights with your fingers but with the pincers or forks provided.

4. *Never place weights on the pans or take weights off, except when the balance is arrested.*

5. The arrestment must be lowered with special care during the first stages of weighing, when the weights on the two sides are not yet nearly equal. Watch the pointer while the arrestment is lowered very slowly. As soon as the pointer is seen to start sharply to one side, raise the arrestment. Notice carefully whether the motion of the pointer to the left or right means that the weights placed in the pan are too small or too great.

6. If the arrestment is to be raised while the balance is swinging, wait till the pointer is nearly at its central position, then raise. The least possible injury will in this way be done to the knife edges.

7. The final weighings must be made with the balance case closed, and care must be taken that the pans do not swing. Large swings of the pans should be carefully stopped by touching the pans with thumb and forefinger or with a camel hair brush while the beam is arrested.

8. In reading successive turning points take no account of the first, which is sometimes irregular.

9. When the weighing is complete, replace the weights carefully into their proper places in the box, remove the arrestment handle, place it in the balance case, and close the case.

Exercise I. To weigh a body using the zero at no load.

1. Find the position of rest of the balance without load. Call this the zero of the balance at no load for the time being.

2. First ascertain by trial on a rough balance, that the weight of the given quartz crystal is say between 100 and 200 grms., then place it in one pan (the left for instance) and weigh to the nearest centigram in the following way: Put 100 grms. in the right pan and add the weight which comes next in descending order of magnitude in the box of weights. Continue adding weights as you have been taught to do in the *Intermediate Course* (pp. 51 and 52), until you find that the addition of another centigram shews excess of weight. Determine the weights on the pan, by noting the empty compartments in the box of weights, and record in your note-book. The number found must be checked by noting the weights themselves as they are removed from the pan at the end of the experiment.

3. Determine the weight to the nearest 2 milligrams.

Use the rider for this purpose, placing it first at the point marked 6 on the beam, then at 8 or at 4, according as the additional 6 mgrms. have been found too small or too great. The student will have been able to proceed so far without taking readings of the pointer. But at this stage he will have to make a rough determination of the position of rest by taking the arithmetical mean of the readings of two consecutive turning points. As he gets nearer to the true value of the weight, he will have to determine the position of rest more accurately, and it will be necessary to read five turning points.

Let it be found in this way, that the weight lies between 185·874 and 185·876 grms.

4. Determine the weight to the tenth part of a milligram.

Observe accurately the positions of rest for the two weights differing by 2 milligrams between which the true value has just been found to lie. Then by interpolation calculate the weight which would bring the position of rest to the zero at no load. With very delicate balances it may be necessary to determine the weight to the nearest milligram before proceeding to interpolation.

5. Again determine the zero at no load.

Enter your results as follows:

Balance: A. Date:

Zero of unloaded balance	10·36
Position of rest with 185·878 grms. in right pan	11·58
„ „ „ „ +2 mgrms. „ „	9·89
Zero of unloaded balance	10·34
Mean zero at no load	10·35
Difference produced by 2 mgrms. 11·58 − 9·89 = ...	1·69
Additional weight required in scale divisions 11·58 − 10·35 =	1·23
„ „ „ in mgrms. $\frac{2 \times 1{\cdot}23}{1{\cdot}69}$ = ...	1·5 mgrs.
Required weight...	185·8795 grams.

Exercise II. To weigh a body by the method of interchanges, sometimes called Gauss's method.

The method given in Exercise I. does not correct for any of the errors of the balance, and if the weight is to be obtained accurately, it is necessary to adopt the method given in the present exercise, or one equivalent to it. The method, which has already been used in Exercise III. of the previous section, consists in interchanging the weights in the pans, and finding directly, or by interpolation, a weight which will bring the balance to the same position of rest, whether the substance to be weighed is in the right or in the left pan. It is clear that two weights which can be interchanged without altering the position of rest of the balance, must be equal.

Proceed as follows:

1. Find the weight to the nearest 2 mgrms., as in Exercise I.
2. Find the position of rest when the substance is placed in the left pan, the lower one of the two limiting weights being placed on the right-hand side of the beam.
3. Interchange weights and substance and find the position of rest, being careful to remove the rider from the right-hand beam, and to place it or a similar one at the corresponding point on the left-hand beam.
4. Interchange once more, so as to bring back the substance into the pan in which it was originally placed.
5. Increase the weight on the right side by 2 mgrms., and determine the position of rest.

Reduce and enter as follows:

Balance: A. Date:

Position of rest with 185·878 grm. weights on right	...	10·57 (*a*)
" " " " left	...	10·00 (*b*)
" " " " right	...	10·45 (*c*)
" " " +2 mgrms. " "	...	8·94
Difference produced by 2 mgrms.	...	1·51
Mean of (*a*) and (*c*)	...	10·51
Zero of balance with load $\frac{10·51 + 10·00}{2}$	...	10·26
Additional weight required in scale divisions 10·26 − 10·00		·26
" " " " mgrms. $\frac{2 \times ·26}{1·51}$ = ...		·34 mgrms.
Required weight	...	185·87834 grams.

Note.—It may happen, if the arms of the balance are not sufficiently equal, that the weight does not in reality lie between the two limits found in Exercise I. If the real weight is *above* the higher limit, the additional weight required to produce the balance will be found greater than 2 mgrms. If, on the other hand, the true weight is *smaller* than the lower limit, the position of rest when the weights are on the right ((*a*) and (*c*) above) gives a lower reading than in (*b*) when the weights are in the left pan. Thus, suppose the reading for (*b*) had been 11·0 instead of 10·0 we should have had to write:

Additional weight required in scale divisions

$$10·26 - 11·00 = -·74,$$

the negative sign indicating that the correction has to be subtracted from 185·878. Students should note carefully whether the correction they find is to be added or subtracted.

It will be observed that the substance is weighed twice in one pan and once in the other. The object of this is to eliminate the effects of changes in the balance which take place if the temperature of the balance case is increasing owing to the approach of the observer, or to the presence of gas flames. The difference between (*a*) and (*c*) in the above example, may either be due to accidental causes or to a systematic change. If the former, the mean will be a more probable value of the position of rest than either of the observed numbers; if the latter, the mean will represent the position of rest at the time at which the observation (*b*) was taken, provided the observations were carried out at nearly equal intervals of time. In any case to get the best result the mean of the first and third observations should be compared with the second. If extreme accuracy is not required, the third observation may be dispensed with; if, on the other hand, it is required to determine the position of rest correctly to the hundredth part of a scale division, it will be necessary to interchange the weights oftener.

To complete the exercise, the result should be checked by placing the rider at the position corresponding to the weight found to the nearest tenth of a milligram (185·8783), and the weighing repeated. The barometer and thermometer should also be read, so that the buoyancy correction may be applied. The barometer need only be read once, unless there is reason to suppose that it is rapidly changing at the time, but the temperature of the balance case should be taken at the beginning and end of the experiment. An example will shew how the final result is now arrived at.

Balance: A. Date:

Assumed weight 185·8783 grs.

Barometer 76·4 cms.

Time 11 h. 15 m. Temperature of balance case ... ... 18°·6 C.
Position of rest with 185·8783 grs. on right ... ... 10·36
" " " left ... ... 10·46
" " " right ... ... 10·39
" " " +2 mgrms. " ... ... 8·91
" " " grs. " ... ... 10·41
11 h. 32 m. Temperature of balance case ... ... 18°·8 C.
Mean position with weights on right (10·36 + 10·39)/2 ... = 10·375
" " " left ... ... 10·46
Zero of loaded balance = (10·375 + 10·46)/2 ... ... = 10·42
Difference produced by 2 mgrms. = 10·40 − 8·91 ... ... = 1·49
Additional weight required in scale divisions 10·42 − 10·46 = −·04

Additional weight required in mgrms. $-\frac{2 \times \cdot 04}{1 \cdot 49}$ ... $= -\cdot 05$ mgrm.

Required weight ... ... 185·8783 − ·00005 = 185·8783 grams.

It will be seen that the repetition of the experiment has led to a result which is the same as that of the previous determination, and if the experiments have been carried out with care, the difference between them should never exceed ·2 mgrm. If the difference exceeds ·25 mgrm. the rider should be placed at the position indicated by the last result, and a fresh determination made.

Buoyancy Correction.

To calculate the buoyancy correction, first calculate by the following method the density of the air, assuming it to be half saturated.

TABLE I.

Density of dry air at ordinary temperatures and pressures.

Temperature	Pressure in cms. of mercury				
	73 cms.	74 cms.	75 cms.	76 cms.	77 cms.
10° C.	·001198	1214	1231	1247	1264
15	1177	1193	1209	1225	1242
20	1157	1173	1189	1204	1220
25	1138	1153	1169	1184	1200

TABLE II.

Maximum pressure of water vapour at ordinary temperatures.

Temperature	Pressure $=p$
10° C.	·91 cms. Hg.
12	1·04 ,,
14	1·19 ,,
16	1·35 ,,
18	1·53 ,,
20	1·74 ,,
22	1·96 ,,
24	2·22 ,,
26	2·50 ,,

Consider a mixture of two gases the densities of which at a pressure P and given temperature are d_1 and d_2. Let the partial pressures of the two gases be p_1 and p_2 respectively, the total pressure being $P = p_1 + p_2$. The density λ of the mixture is

$$\lambda = \frac{d_1 p_1}{P} + \frac{d_2 p_2}{P} = \frac{d_1}{P}\left\{P + \frac{d_2 - d_1}{d_1} p_2\right\},$$

i.e., the density of the mixture is equal to that which the first constituent alone would have at the pressure

$$P + \frac{d_2 - d_1}{d_1} p_2 .$$

If the two gases are air and aqueous vapour respectively, the index 2 referring to the latter, the ratio d_2/d_1 is very nearly 5/8, hence

$$P + \frac{d_2 - d_1}{d_1} p_2 = P - \frac{3}{8} p_2 .$$

If then p_2, the pressure of the aqueous vapour present in the atmosphere at the time, is known, we may use Table I. to determine the density λ, by taking the air to be dry at a pressure $P - \frac{3}{8} p_2$, instead of saturated at the observed barometric pressure P, and if we assume the air to be half saturated, *i.e.*, take $p_2 = \frac{p}{2}$, p being the maximum pressure possible at the observed temperature, given in Table II., we should have to take the equivalent air pressure to be $P - \frac{3}{16} p$. The value of p at the temperature of the room will on the average be about 15 mm.; the error we should make if the air happened to be totally dry or totally moist, would therefore be the same as if we had measured the height of the barometer incorrectly by about 3 mm., which would cause an error in the density of the air and in the buoyancy correction, of about one part in 250. If we require to make the correction with certainty to less than that amount, we should have to measure the pressure of aqueous vapour in the balance case.

In the above example the barometer stood at 76·38 cms., and the mean temperature of the balance case was 18°·7 C.

Pressure of aqueous vapour at 18°·7 C.... ... = 1·6 cms.

$\left(\frac{3}{16}\right) p$ ... ... = ·3 „

P ... ... = 76·38 „

$P - \frac{3}{16} p$ (to the nearest millimetre) ... ... = 76·1 „

Density of dry air at 18° C. and 76·1 cms. pressure (by interpolation) from Table I. ... = ·001215

Density of dry air at 19° C. and 76·1 cms. pressure (by interpolation) from Table I. ... = ·001211

Density of dry air at 18°·7 C. and 76·1 cms. pressure (by interpolation) from last two values = ·001212

To find the buoyancy correction $\lambda \left[\frac{W}{\rho} - \frac{W_1}{\sigma_1} - \frac{W_2}{\sigma_2}\right]$ the specific gravity of the body weighed must be known approximately. In the above example the body was Quartz, so that if W and ρ refer to the body weighed, W_1 and σ_1 to the brass weights, and W_2, σ_2 to the platinum weights,

$$\frac{W}{\rho} = \frac{185{\cdot}88}{2{\cdot}653} \quad \ldots \quad = 70{\cdot}06$$

$$\left.\begin{aligned} \frac{W_1}{\sigma_1} &= \frac{185}{8{\cdot}40} \quad \ldots \quad = 22{\cdot}02 \\ \frac{W_2}{\sigma_2} &= \frac{{\cdot}88}{21{\cdot}5} \quad \ldots \quad = {\cdot}04 \end{aligned}\right\} = 22{\cdot}06$$

$$\frac{W}{\rho} - \frac{W_1}{\sigma_1} - \frac{W_2}{\sigma_2} \quad \ldots \quad = 48{\cdot}00$$

∴ buoyancy correction = ·001212 × 48·00 = ·05817

Weight found... ... ... = 185·87825

Weight corrected (to the nearest tenth of a mgrm.) ... ... = 185·9364 grams.

It will be seen that if we had neglected the platinum weights, we should have made an error of about 1/20 mgrm., which, considering the uncertainty in the assumed specific gravity of the brass weights used, would have been allowable, but if we had assumed *all* the weights 185·88 to be brass, the error would have amounted to more than the tenth of a milligram. This example shews how difficult it is to obtain a weight correctly to one part in a million.

Numerical Exercise.

A litre of water is to be weighed to the nearest milligram; calculate how nearly you require to know the height of the barometer, the temperature of the balance case, and the specific gravity of the brass weights.

Exercise III. To standardise a box of weights.

(*a*) In terms of the unit of the box.

If the box has 3 weights of unit magnitude select one (1) as the unit and call the others 1′ and 1″. Weigh each weight against smaller weights making up the same magnitude by Gauss's method, supply the corrections for buoyancy so as to obtain the following equations:

$$\begin{aligned}
1' &= 1 + a, \\
1'' &= 1 + b, \\
2 &= 1 + 1' + c = 2.1 + a + c, \\
5 &= 2 + 1 + 1' + 1'' + d = 5.1 + 2a + b + c, \\
&\text{\&c.},
\end{aligned}$$

where a, b, c, &c. are small quantities positive or negative.

(*b*) In terms of any weight N of the box which may have been compared with a standard weight.

The previous observations give

$$N = n.1 + k,$$

hence $1 = \dfrac{N - k}{n}$, and since N is known in terms of the standard the unit of the box, and therefore each weight, is given in terms of the standard.

SECTION VIII

DENSITY OF A SOLID

Apparatus required: *Delicate balance, piece of quartz, specific gravity flask, air and water baths, small pieces of quartz.*

The density of a substance at any point is defined to be the quotient of the mass of a small volume of the substance at that point, by the volume. If the substance is homogeneous and m is the total mass, and v the total volume, we have $\rho = m/v$.

Hence if the unit of mass is the gram, and the unit of length the centimetre, the density of a homogeneous body will be numerically equal to the mass of one cubic centimetre of the substance.

In the metric system, the gram was originally chosen to be the mass of 1 cubic centimetre of water at its point of maximum density 3°·95 C. A kilogram, equal to 1000 grams, deposited in Paris, serves as the ultimate standard for weights constructed on the metric system. Since the experiments determining the gram were made, however, physical instruments and methods of observation have improved, so that a small difference is now found to exist between the theoretical gram and the practical standard of mass. In consequence, the density of water at 3°·95 C. is not unity, as it was meant to be, but is 1·000013. The difference is so small that it may generally be neglected, but it would have to be taken into account if, for instance, 100 grams of water were to be weighed correctly to a milligram, and the volume occupied by the water calculated from the results to one part in 100,000.

The specific gravity of a homogeneous substance is defined to be the ratio of the mass of any volume of the substance, to the mass of the same volume of water at

3°·95 C. As we may take the mass of 1 c.c. of water at 3°·95 C. to be unity, it is clear that the specific gravity is numerically equal to the density expressed in the C.G.S. system of units. This is an advantage, as in that system we may dispense altogether with the idea of "specific gravity" and always use that of "density" instead. Students must be clear, however, that the two terms are not synonymous, as the numerical value of the specific gravity is independent of the units of length and mass, while the number expressing the density will depend on those units.

Since the direct determination of the volume of a body cannot be carried out accurately unless the body is of some regular shape, density determinations generally depend on a previous knowledge of the density of some standard substance, water being selected as the most convenient standard.

Increase of temperature will in general diminish the density of a body, hence in stating the density, the temperature at which the number holds should always be specified.

The density of water has been determined with great care, and has been found to decrease at an increasing rate per degree as the temperature rises. At 15° C., the ordinary temperature of the laboratory, the decrease of density of water for 1° C. is not much more than 15 parts in 100,000, while at 50° C. it is 5 parts in 1000. Tables giving the densities at various temperatures are given in Kaye and Laby's *Physical Tables*, and Landolt and Börnstein's *Physikalisch-Chemische Tabellen*, but the Table on page 62 will be sufficient for most purposes.

There are a number of different methods by means of which the density of a body may be determined, and the best manner of proceeding in each case will depend on the available quantity of the substance, on its chemical properties and state of aggregation. We shall describe two methods, one of which will always be applicable if the substance is solid and insoluble in water. The method would have to be modified if the substance were soluble in water, or if it were hygroscopic.

Method I. If a body is weighed first in air, and then suspended in a liquid, its apparent weight will be less in the second case than in the first, and it has been known since the time of Archimedes, that the apparent loss of weight is equal to the weight of the liquid displaced by the body. If M is the mass of the body, and ρ its density, its volume is M/ρ, and if σ is the density of the fluid in which it is weighed, the apparent decrease of mass will be

$$M\frac{\sigma}{\rho}.$$

The same holds for a weighing in air, the density λ of air being substituted for σ.

Assuming the arms of the balance to be a and b cms. respectively, we have for the moments about the central knife-edge, of the forces on the two arms during the weighing in air, the quantities

$$aMg\left(1-\frac{\lambda}{\rho}\right) \text{ and } bM_1g\left(1-\frac{\lambda}{\sigma_1}\right),$$

where M_1 is the apparent mass, and σ_1 the density of the weights. For equilibrium these moments must be equal, hence

$$aM\left(1-\frac{\lambda}{\rho}\right)=bM_1\left(1-\frac{\lambda}{\sigma_1}\right).$$

Similarly for the weighing in water, if M_2 is the apparent mass

$$aM\left(1-\frac{\sigma}{\rho}\right)=bM_2\left(1-\frac{\lambda}{\sigma_1}\right).$$

Dividing the first equation by the difference between the first and second, we have

$$\frac{\rho}{\sigma-\lambda}\cdot\left(1-\frac{\lambda}{\rho}\right)=\frac{M_1}{M_1-M_2}.$$

Or

$$\rho-\lambda=\frac{M_1}{M_1-M_2}(\sigma-\lambda),$$

i.e., the excess of density of body over that of air

$$=\frac{M_1}{M_1-M_2}\times \text{(excess of density of liquid over that of air).}$$

It will be noticed that neither the inequality of the arms of the balance, nor the buoyancy effect of the air on the brass weights, enters into the result, so long as the weights are always placed in the same pan. This is due to density determinations depending on ratios of weights only.

It will also be seen that the equation giving ρ corrected for the buoyancy of the air, may be obtained from the equation

$$\rho = \frac{M_1}{M_1 - M_2} . \sigma,$$

in which the air is neglected, by subtracting the density of the air from each density occurring in the equation. This may be seen on consideration to be due to the fact, that the weight of a body obtained in air would be the same as the weight obtained *in vacuo*, if the density of the body as it was transferred from air to vacuo, were decreased by the density of the air. This holds for both the quartz and the water in the above case, and the principle will be used in other cases.

The numerical calculation is best carried out by writing $1 - \kappa$ for σ, and ρ' for $M_1/(M_1 - M_2)$, when we have

$$\rho = \rho' - \rho' (\kappa + \lambda) + \lambda,$$

where the last two terms are small.

Exercise I. Determination of the density of Quartz by weighing in water.

1. Place a wooden stool across the left-hand pan, place upon it an empty beaker of suitable size. Estimate the length l between the hook at top of the balance pan and the centre of the beaker, and take two pieces of silk about 20 cms. longer than this estimated length, use one to tie round the quartz crystal provided, leaving a length l with a loop at the end, hanging from the crystal. Cut away all unnecessary thread, and cut off equal lengths from the other piece.

2. Suspend the quartz by means of the thread from the hook underneath the top of the support of the left-hand balance pan, place the other piece of thread in the right-hand pan, and find the weights required to produce equilibrium.

3. Remove the quartz from the balance, place it in water in a beaker, and boil the water to drive off the air bubbles adhering to the quartz. Then cool the water by pouring into it water from the tap gently without causing splashes. Place a small wooden stool across the left-hand pan of the balance, so that the pan does not come into contact with it at any point. Support the beaker on this stool and suspend the quartz again from the hook. See that the quartz is entirely immersed in the water and that no bubbles of air adhere to it. Place a thermometer in the water, and note the temperature (Fig. 16). Cut off from the piece of thread in the right-hand pan a length equal to the length of thread in the water, and remove it. Weigh the quartz.

Record as follows:

Density of Quartz. Method I.

Date:

Balance A. Box of Weights A.

Weights in right-hand pan throughout.

Temperature of Balance Case, 18°·4 C.

Weight of Quartz in air (M_1) ... ... 40·882 grms.

Temperature of Water, 20° C.

Apparent Weight of Quartz in Water (M_2) ... 25·487 ,,

Loss of Weight = 15·395 ,,

$$\rho' = \frac{M_1}{M_1 - M_2} = \frac{40{\cdot}882}{15{\cdot}394} = 2{\cdot}6556$$

$$\sigma = {\cdot}9983$$

$$\lambda = {\cdot}0012$$

$$\sigma - \lambda = {\cdot}9971$$

$$\frac{M_1}{M_1 - M_2}(\sigma - \lambda) = 2{\cdot}6479$$

$$\lambda = {\cdot}0012$$

$$\rho = 2{\cdot}6491$$

Hence ρ the density of the quartz at 20° C. = 2·6491.

Method II. If the solid can only be obtained in small pieces, we may determine the density by the use of a "specific gravity flask" (Fig. 19), which is a small glass flask provided with a well-ground stopper traversed by a narrow channel. When it is filled with a liquid, and the stopper is inserted carefully so as to exclude air bubbles, the excess of liquid will flow out through the capillary opening. By means of a piece of blotting-paper a small quantity of the liquid may be removed, so that it just reaches to a marked height in the opening. The flask may in this way be repeatedly filled to the same level, and if its temperature is the same, the volume of its contents will be the same.

FIG. 19

The flask having been cleaned and dried, the density required is determined by the following series of weighings, the letters, F, &c., representing the weights obtained:

1. The flask dry, F.
2. The flask dry with the dry solid placed inside, $F + M_1$.
3. The flask with the solid inside, after filling up to the mark with a liquid of known density σ_1 at a temperature t_1, $F + M_1 + W_1$.
4. The flask entirely filled up to the mark with a liquid of density σ_2 at a temperature t_2, $F + W_2$.

Since σ_2 is the density of the liquid at t_2, the volume of the flask at t_2, neglecting the effect of the air on the weighing, is $\frac{W_2}{\sigma_2}$, and the volume at t_1 will be $\frac{W_2}{\sigma_2}\frac{1 + \alpha t_1}{1 + \alpha t_2}$, where α is the coefficient of cubical expansion of glass. Similarly the volume occupied by the liquid at $t_1 = \frac{W_1}{\sigma_1}$. The difference between these volumes is the volume occupied by the solid at t_1, and the density ρ of the solid at t_1 is the mass divided by this volume.

The effect of the buoyancy of the air may be taken into account by subtracting the density of air λ, from each of the densities in the equation for ρ (page 56), and we thus get

$$\rho - \lambda = \frac{M_1}{\frac{W_2}{\sigma_2 - \lambda}\cdot\frac{1 + at_1}{1 + at_2} - \frac{W_1}{\sigma_1 - \lambda}}.$$

If we take weighings of the flask empty and when filled with each of two liquids, we may compare the densities of the liquids, since on making $M_1 = 0$ in the above equation, we have

$$\frac{W_2}{\sigma_2 - \lambda}\cdot\frac{1 + at_1}{1 + at_2} = \frac{W_1}{\sigma_1 - \lambda},$$

an equation from which we can determine one density if the other is known.

If in the former equation, we take the liquid to be water in each case, so that σ_1 and σ_2 are both nearly unity, we have, using the methods of approximation given in *Intermediate Practical Physics* (page 16),

$$\rho - \lambda = \frac{M_1}{W_2\,(1 + \overline{\sigma_1 - \sigma_2} + a\overline{t_1 - t_2}) - W_1}\,.(\sigma_1 - \lambda).$$

Or writing $\sigma_1 = 1 - \kappa$, and

$$\rho' = \frac{M_1}{W_2\,(1 + \overline{\sigma_1 - \sigma_2} + a\overline{t_1 - t_2}) - W_1},$$

we have

$$\rho = \rho' - \rho'\,(\kappa_1 + \lambda) + \lambda.$$

Exercise II. Determination of the density of Quartz by the specific gravity flask method.

1. Clean the 50 gram flask (Fig. 19) provided, by washing it if necessary with a strong solution of caustic potash, and then with water from the tap. The potash is to be thoroughly removed with tap water, and the final washing made with distilled water.

Place the flask on the shelf of an air bath kept at about 120° C., insert a glass tube into the flask, through an opening in the top of the bath. Look at the flask occasionally, and when the drops of water have evaporated from the sides, draw the moist air out of the flask, by applying the mouth to the upper end of the glass tube. Repeat this several times, then remove the flask and allow it to cool. If no moisture is deposited on its inside surface as it cools, place it in the left-hand pan of the balance, and weigh.

With the quantities of the substance available, it will be sufficient to weigh accurately to the nearest milligram, and the process of weighing may therefore be shortened by taking only three readings of the vibrating pointer (two on one side and one on the other), instead of five, as in the previous exercises.

2. Dry thoroughly 15 or 20 grams of the broken up pieces of quartz with which you are provided, and place in the flask. Weigh the flask and contents, and hence deduce the weight of the quartz.

3. Pour some distilled water into the flask. If air bubbles adhere to the small pieces of quartz, it will be necessary to expel them. For this purpose place the flask, with the solid pieces completely covered with water, in the air bath, and heat up carefully until the water just begins to boil. Take out the flask, and after allowing a little time for cooling, fill up with distilled water, and then place it in the water bath with which you are provided, keeping it in position by india-rubber bands.

Stir the water well, and keep it at a temperature two or three degrees higher than that of the room.

Read the temperature to $0^\circ{\cdot}05$ C.

The temperature of the bath must be kept constant, and the flask kept in the bath for a sufficient length of time to allow the water in it to take up the same temperature. This may be ascertained by placing a thermometer in the flask.

The stopper is then inserted, and the quantity of water in the flask is adjusted so that it reaches to the mark across the capillary opening in the stopper.

Take the flask out of the water and dry the outside carefully, taking care to avoid heating it by contact with the hand and forcing out any of the water. The object of filling the flask with water at a temperature above that of the air is now apparent, for when the flask is taken out of the water bath, the water in it will contract, and thus the danger of losing any by accidental heating of the flask is greatly diminished. After the flask has been dried on the outside, leave it in the balance case for a few minutes so that it may acquire the temperature of the balance case, then weigh again.

If the flask is not perfectly dry outside, evaporation will take place and the weight of the flask will slowly diminish. Ascertain that the weight remains constant.

The difference between the weights of the flask in this and the previous weighing will give the weight W_1 of the water it contains.

4. Now take out the quartz, fill the flask with distilled water and place it in the water bath. Keep the temperature of the bath nearly the same as it was in the second weighing; a correction will be necessary if any difference exists between the two temperatures, which should be known to the twentieth of a degree. Close the flask, dry and weigh it.

Record as follows:

Density of Quartz. Method II.
Date:

Balance *A*. Weights *A*. Flask 13.
Weights in right-hand pan throughout.
Temperature of Balance Case, 16°·0 C.

Weight of flask, F	=	17·325	grms.
Weight of flask and quartz, $F + M_1$ ...	=	32·012	,,
Hence weight of quartz, M_1	=	14·687	,,
Weight of flask, quartz, and water at $t_1 = 18°$, $F + M_1 + W_1$ ...	=	76·438	,,
Hence weight of water at 18°, W_1 ...	=	44·426	,,
Weight of flask filled with water at $t_2 = 19°$*, $F + W_2$ ...	=	67·274	,,
Hence weight of water filling flask at 19°, W_2 ...	=	49·949	,,

* This temperature differs more from t_1 than it need be allowed to do. The difference is taken great here to shew clearly the magnitude of the correction introduced.

$$
\begin{array}{lcl}
\text{At } 18^\circ,\ \sigma_1 & = & \cdot 99862 \\
\text{At } 19^\circ,\ \sigma_2 & = & \cdot 99842 \\
\sigma_1 - \sigma_2 & = & \cdot 00020 \qquad \alpha = \cdot 000023 \\
\alpha\,(t_1 - t_2) & = & -\cdot 00002 \\
\text{Sum} & = & \cdot 00018
\end{array}
$$

$$
\begin{array}{lcl}
W_2\,(\sigma_1 - \sigma_2 + \alpha \overline{t_1 - t_2}) \;\ldots\;\ldots\;\ldots & = & \cdot 009 \text{ grms.} \\
W_2\,(1 + \sigma_1 - \sigma_2 + \alpha \overline{t_1 - t_2}) \;\ldots\;\ldots & = & 49\cdot 958 \;\;\text{,,} \\
W_1 \;\ldots\;\ldots\;\ldots\;\ldots\;\ldots & = & 44\cdot 426 \;\;\text{,,} \\
W_2\,(1 + \sigma_1 - \sigma_2 + \alpha \overline{t_1 - t_2}) - W_1 \;\ldots & = & 5\cdot 532 \;\;\text{,,} \\
\rho' = \dfrac{14\cdot 687}{5\cdot 532} \;\ldots\;\ldots\;\ldots\;\ldots & = & 2\cdot 6550
\end{array}
$$

$$
\begin{array}{rcl}
\kappa_1 & = & \cdot 00133 \\
\lambda & = & \cdot 00122 \\
\kappa_1 + \lambda & = & \cdot 00255
\end{array}
$$

$$
\begin{array}{lcl}
-\rho_1\,(\kappa_1 + \lambda) \;\ldots\;\ldots\;\ldots\;\ldots & = & -\cdot 0067 \\
\lambda \;\ldots\;\ldots\;\ldots\;\ldots\;\ldots & = & \cdot 0012
\end{array}
$$

Hence density ρ of the quartz at 18° C. $= 2\cdot 6495$

TABLE OF THE DENSITY OF WATER.

Temp.	Density
0° C.	0·99987
5°	·99999
10°	·99973
15°	·99913
20°	·99823
25°	·99707
30°	·99567

SECTION IX

DENSITY OF A LIQUID

Apparatus required: *Balance, specific gravity flask, salt solution, water bath, Mohr's balance, and hydrometer.*

Method I. By the specific gravity flask.

In the previous section (p. 59) it has been shewn how the density of a liquid may be found by means of the specific gravity flask, and we now proceed to apply the method to the determination of the density of the salt solution provided.

Make use of the specific gravity flask used in the determination of the density of quartz. The apparent weight of the flask empty was found to be F grams, and the apparent weight of water filling it at the temperature t_1, W_1 grams.

The flask should be dried, filled with salt solution, and placed in the water bath at a temperature 2 or 3 degrees higher than that of the balance case.

After allowing the solution to take up the temperature of the bath, put in the stopper and carefully remove with filter paper the drop at the top of the hole in the stopper. Let the observed temperature of the water bath be t_2. Remove the flask, dry the outside and weigh. Let W_2 be the apparent weight of the salt solution.

The volume of the flask at the temperature t_1 is $\frac{W_1}{\sigma_1}$ when σ_1 is the density of water at that temperature; the volume of the flask at 0° C. will therefore be $\frac{W_1}{\sigma_1}\Big/(1 + \alpha t_1)$, where α is the coefficient of expansion of glass, and this volume must be the same as that calculated from the corresponding weighing of the liquid, *i.e.* $\frac{W_2}{\sigma_2}\Big/(1 + \alpha t_2)$. The correction of the weighings for buoyancy is introduced by writing $\sigma_2 - \lambda$ for σ_2 and $\sigma_1 - \lambda$ for σ_1 (see page 56), hence

$$\frac{W_2}{\sigma_2 - \lambda} = \frac{1 + at_2}{1 + at_1} \cdot \frac{W_1}{\sigma_1 - \lambda}.$$

Or since at_1, at_2 are both small

$$\sigma_2 - \lambda = \frac{W_2}{W_1}(1 + a\overline{t_1 - t_2})(\sigma_1 - \lambda),$$

which equation serves to calculate σ_2 if σ_1 is known.

Record as follows:

Density of Salt Solution. Method I.

Date:

Balance *A*. Weights *A*. Flask 13.

Weights in right-hand pan throughout.

Temperature of Balance Case, 18° C.

Weight of dry flask, F	= 17·325	grms.
,, ,, flask filled with water at 19°	= 67·274	,,
,, ,, water filling flask at 19° = W_1	= 49·949	,,
,, ,, flask filled with salt solution at 19°·2	= 70·282	,,
,, ,, salt solution filling flask at 19°·2 = W_2	= 52·957	,,

Density of water at 19° = σ_1	=	·9984
,, ,, air at 18° = λ	=	·0012
$\therefore \sigma_1 - \lambda$ ...	=	·9972

a for glass = ·000023, $\therefore a\overline{t_1 - t_2} = 0$.

$\frac{W_2}{W_1} = \frac{52·957}{49·949}$	=	1·0601
$\sigma_2 - \lambda = 1·0601 \times ·9972$ ...	=	1·0572
λ	=	·0012
$\therefore \sigma_2$ = density of solution	=	1·0584

Method II. By Mohr's Balance.

Mohr's balance (Fig. 20) is an ordinary balance modified so as to enable determinations of densities of liquids to be made rapidly. One arm of the balance is divided into 10 equal parts, and carries, suspended from its end by a fine silk fibre, a glass thermometer which is immersed in and indicates the temperature of the liquid. The other arm carries a counterpoise, the end of which is pointed, and comes close to a corresponding pointer on the frame of the balance. The balance can be clamped at any height by means of the screw in the stem.

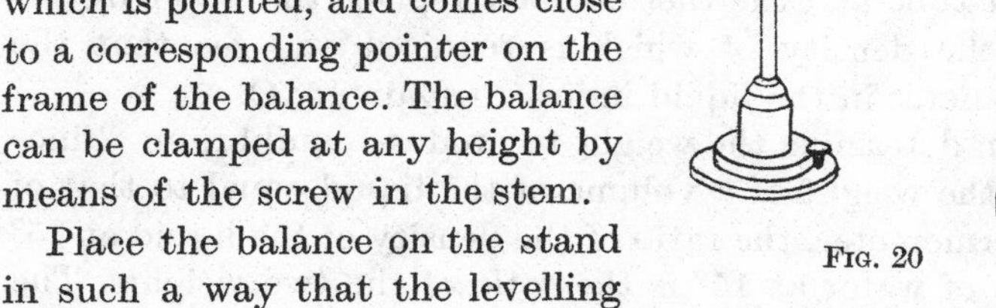

FIG. 20

Place the balance on the stand in such a way that the levelling screw in the stand lies in the vertical plane through the beam.

Suspend the thermometer from the graduated arm, and turn the levelling screw till the pointers are in line with each other. The compensating weight above the thermometer is generally so adjusted that this is possible, but if the pointers cannot be brought into line weights may be hung on the hook from which the thermometer is suspended. Fill the small test tube with water at 15° C. Raise the balance by means of the expanding stem, place the test tube under the thermometer, and lower the balance till the thermometer is entirely immersed in the water. The balance will no longer be in equilibrium owing to the upward force of the water on the thermometer, which we have seen is equal to the weight of the water displaced by the thermometer. To produce equilibrium weights must be placed in the notches of the arm of the balance.

The largest brass weights provided are equal, the other weights are $\frac{1}{10}$ and $\frac{1}{100}$ respectively of the largest weights.

The notches are marked from the centre of the arm to the end 1, 2, 3......9, the hook under the end of the arm being at the tenth notch. If we call the weight of the largest brass pieces 1, then if 1 is placed in the notch 6 it is equivalent to a weight of ·6 placed at the end, and so on. Thus the weight necessary to produce equilibrium is given by the readings of the notches in which the weights in order of magnitude are placed, the unit in which the weight is expressed being that of the largest brass weight. The weight found is in terms of this unit, *i.e.* of a volume of water at 15° equal to the volume of the thermometer.

Remove the weights, raise the balance, and after drying the test tube and the thermometer replace the water by the liquid the density of which is required and see that the thermometer in the liquid indicates again 15° C.

Again determine the weight to produce equilibrium. Since this is the weight of a volume of the liquid equal to that of the thermometer, the ratio of the density of the liquid at 15° to that of water at 15° is the ratio of the two weights. The density of the liquid at 15° is therefore the product of this ratio and the density of water at 15° C. (= ·999).

The weights are generally arranged so that one of the heaviest will when hung in the hook produce equilibrium when the thermometer is in water. In this case the density of the liquid can, to within ·001, be read off immediately from the positions of the weights when the thermometer is in the liquid.

Method III. By the Hydrometer.

When a solid floats on a liquid it is acted on by two forces, one its weight downwards, and the other the pressure of the liquid upwards. Since the body is in equilibrium these two forces must be equal. But the pressure of the liquid is equal to the weight of the liquid displaced. Hence if W be the mass of the solid, V the volume of liquid displaced, σ its density, we must have $W = V\sigma$, or

$$V = \frac{W}{\sigma}.$$

Hence when the solid floats on a denser liquid it sinks less than it does in a lighter liquid and the volume immersed varies inversely as the density of the liquid.

The Hydrometer provided (Fig. 21) has been graduated by being placed in liquids the densities of which had been determined by the previous methods. Hence to get the density of a liquid it is simply necessary to place the hydrometer in it and read the position of the surface of the liquid on the graduated stem.

A test tube mounted on a block of wood is provided for holding the liquid. To read the scale, hold the eye below the level of the surface of the liquid, and gradually raise it till that surface is seen foreshortened into a straight line. The position of this line on the scale is the density required.

Collect your results as follows:

Density of salt solution by	flask	1·0584
,, ,, ,,	Mohr's balance ...	1·059
,, ,, ,,	hydrometer ...	1·058

Fig. 21

SECTION X

MOMENTS OF INERTIA

Apparatus required: *Rectangular metal block, fine wire bifilar supports.*

The moment of inertia of a body about an axis through its centre of gravity may be calculated from its mass and dimensions if the body is regular in form and of uniform density, and if not it may be found experimentally by suspending the body, either by a single fibre or by a double one, in such a way that it can perform torsional oscillations about that axis, and determining the time of an oscillation. If the torsional constant of the suspension is known this enables the moment of inertia to be calculated, but if it is unknown a second body, the moment of inertia of which about some axis through its centre of gravity is known, either by calculation if it is of regular shape or by previous determination, is then attached to the first, in such a way that its axis is coincident with that of the suspension, and the time of an oscillation again determined. From these times the moment of inertia of the first body can be found in terms of that of the second by the equation: $I_0 = I_1 T_0^2/\overline{T_1^2 - T_0^2}$, where I_1 is the moment of inertia of the mass added, T_0 and T_1 the times of oscillation. This is the method often used when the moment of inertia of a suspended magnet is to be determined.

If the bifilar method of suspension is adopted, the body, the moment of which is required, is laid in a stirrup at the lower end of the suspension, with the axis about which the moment is required vertical and half-way between the two suspending fibres. If it is rotated in a horizontal plane through a small angle, the suspension resists the rotation with a force proportional to the sine of the angle of twist.

The couple N for small angles of rotation being $= Mg\,\dfrac{d_1 d_2}{4l}$, we obtain the time of oscillation by substituting this value in the general equation for torsional oscillations

$$T = 2\pi\sqrt{\frac{I}{N}},$$

where M = mass of body suspended.
I = moment of inertia of body suspended about axis of suspension.
l = vertical distance between ends of fibres.
d_1, d_2 = distances of fibres apart at top and bottom.
g = gravitational acceleration.

Since $I = Mk^2$, where k is the radius of gyration of the body about the axis, we have for a small oscillation

$$T = 2\pi\sqrt{\frac{4lk^2}{gd_1 d_2}}.$$

Exercise I. To verify the relation between the time of oscillation and the constants of the bifilar suspension.

Suspend from the lower end of the fibres the rectangular block provided (Fig. 22), passing the fibre through the holes at the ends of the small brass strip which can be screwed to the block, then over the larger pulley above the separated horizontal clamps forming the upper support. When the block has come to rest, screw up the clamps so that the two fibres are held firmly, allow the block to come to rest, place the vertical wire pointer close to the edge of the block, then set the block in oscillation through an arc of about 20°, and determine the time of oscillation (Section III), and the mean arc of twist on each side of the equilibrium position by the method described, p. 14. Measure the vertical distance between the ends of the fibres and their distances apart at the top and bottom.

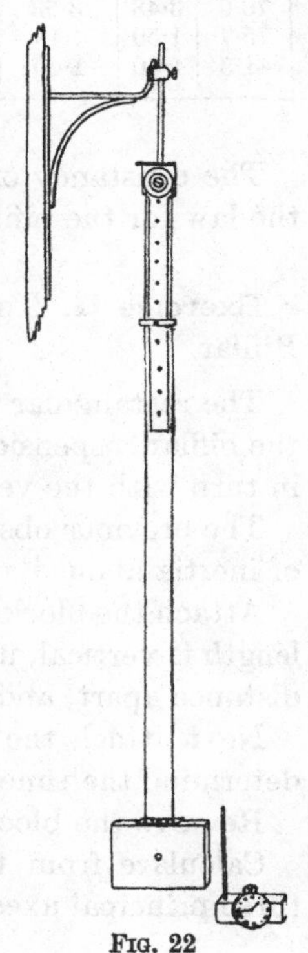

FIG. 22

Now unscrew the clamp, pass the fibre

through the holes nearer the centre of the strip, and over the smaller pulley, then clamp it, again determine the time of oscillation, measure again the length l and the distances apart at top and bottom.

Reduce the length of the fibres by moving the clamp to a lower part of the support and again clamping the fibres. Again determine the time, and measure the length l and distances apart of the fibres.

Tabulate the observations and results as follows:

Apparatus *A*. Date:

Length cms.	Distances apart cms.		$d_1 d_2$	$\frac{d_1 d_2}{l}$	Axis of block vertical	Time seconds	(time)^2	$\frac{d_1 d_2}{l} t^2$
	Top	Bottom						
76·0	3·48	3·53	12·28	·161	longest	2·927	8·57	1·38
75·7	1·50	1·40	2·10	·0277	,,	6·960	48·35	1·34
41·5	1·50	1·40	2·10	·0506	,,	5·201	27·00	1·37

The constancy of the numbers in the last column verifies the law for the bifilar suspension.

Exercise II. To determine a Moment of Inertia by the Bifilar.

The rectangular metal block provided may be attached to the bifilar suspension, so that its three principal axes coincide in turn with the vertical axis of the suspension.

The previous observations suffice to determine the moment of inertia about its longest axis.

Attach the block to the suspension so that the axis of mean length is vertical, using the full length of fibre and maximum distance apart, and determine the time of oscillation.

Next attach the block with its shortest axis vertical and determine the time.

Remove the block and weigh it.

Calculate from the observation the moments about the three principal axes, arranging the work as follows:

Block marked A. Date:

Weight of block 1200 grams.

$$l = 76\cdot 0 \text{ cms.}$$

$$\left.\begin{matrix} d_1 = 3\cdot 48 \\ d_2 = 3\cdot 53 \end{matrix}\right\} \therefore \frac{d_1 d_2}{l} = \cdot 161, \quad \frac{g d_1 d_2}{4l} = 39\cdot 5.$$

Axis vertical	Time sec.	t^2	$\frac{t^2}{4\pi^2}$	$\frac{t^2}{4\pi^2}\frac{gd_1d_2}{4l} = k^2$	I observed	I calculated
long	2·927	8·57	·215	8·64	10360	10100
mean	4·33	18·49	·463	18·61	22330	22100
short	5·128	26·30	·659	24·49	31800	32400

The fifth column gives the values of k^2.

Determine the dimensions of the block, and calculate the moments by the formula

$$I_a = M\frac{b^2 + c^2}{3},$$

where a, b, c are the half lengths of the sides of the block.

Record as follows and enter in preceding table.

$$M = 1200 \text{ grams}$$

$$a = 7\cdot 45 \text{ cms.} \therefore a^2 = 55\cdot 5 \therefore I_a = 10100$$

$$b = 5\cdot 05 \qquad b^2 = 25\cdot 5 \qquad I_b = 22100$$

$$c = \cdot 50 \qquad c^2 = \cdot 25 \qquad I_c = 32400$$

The oscillations of a bifilar suspension like those of a pendulum are only approximately isochronous, and the arc of rotation should therefore be small if great accuracy is required. If the arcs are large the correction given in the next exercise may be applied.

The following modification of Routh's rule for writing down the Moment of Inertia of a symmetrical body—a brick, a cylinder or an ellipsoid—about an axis through its centre of mass is useful. Let M be the mass of the body, a and b its semi-axes perpendicular to that about which the moment is required, C_a, C_b the type of curvature of the surface of the body in which each axis a, b ends (single, $C = 1$, or double, $C = 2$), then

$$I = M\left\{\frac{a^2}{3 + C_a} + \frac{b^2}{3 + C_b}\right\}.$$

Exercise III. To determine the Moment of Inertia of an irregular body about an axis through its centre of gravity by the Unifilar. Attach the body to the unifilar with the axis about which the moment I_0 is required in line with the unifilar. Determine the time of a torsional oscillation T_0. Attach to the body a second body whose moment of inertia I_1 about the axis of the suspension is known and determine the time of oscillation T_1. Calculate I_0 by the equation of the opening paragraph.

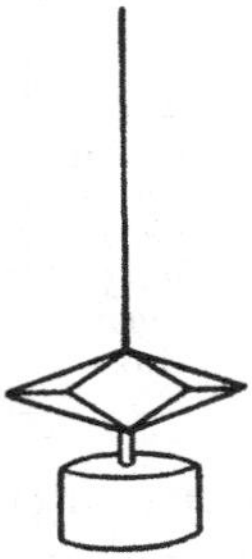

FIG. 23

SECTION XI

THE GRAVITATIONAL ACCELERATION BY THE REVERSIBLE PENDULUM

Apparatus required: *Brass bar with holes, knife edge on adjustable support, and clock or chronometer.*

A brass bar, in which a number of holes have been drilled in order to make it unsymmetrical, is provided (Fig. 24). It is supported on a knife edge projecting from a plate attached to the wall by screws which allow the knife edge to be levelled (Fig. 25). On hanging the bar on the knife edge by one of the holes in it, it can be set swinging about an axis at right angles to its length at a distance from its centre of gravity which can be varied.

Fig. 24

See that when the pendulum is pulled aside and released it swings in a plane without rotating about its own axis on the knife edge. This can be secured by levelling the knife edge.

Fig. 25

Place the end hole of the lighter end of the bar on the knife edge and determine the time of a small oscillation not exceeding 6° on each side of the vertical with the help of a clock beating seconds or a chronometer beating half-seconds, as described on p. 14.

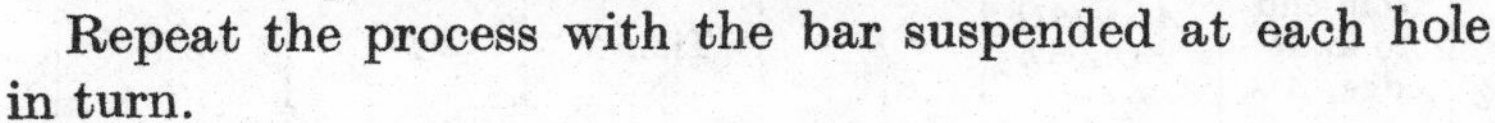

Repeat the process with the bar suspended at each hole in turn.

Plot a curve (Fig. 26) with distances of the point of suspension from the heavier end of the bar as abscissae and the square of the time of oscillation as ordinates.

Determine the centre of mass G of the bar by balancing it on a knife edge, measure its distance from the heavier end of the bar and mark the point on the diagram. Note that the curve is symmetrical about the ordinate through G.

From the measurements of the positions of the knife edge

O and of the centre of mass G determine the distances $OG = h$, of the centre of mass below the knife edge (Fig. 27),

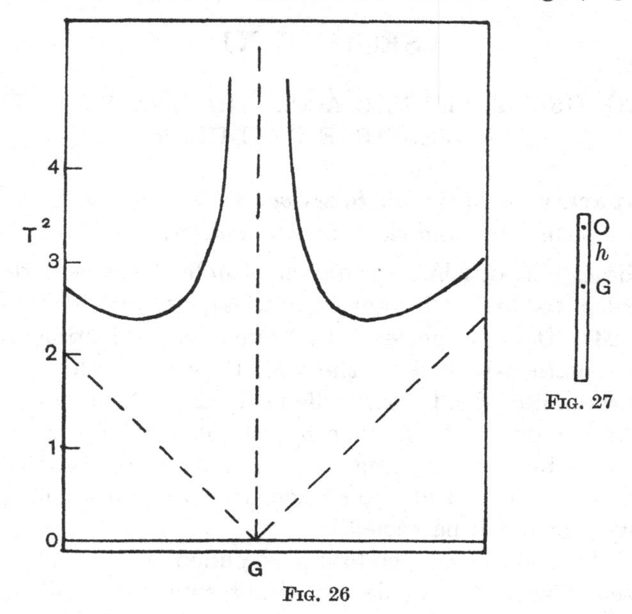

Fig. 27

Fig. 26

find the values of h^2 and hT^2 for each position and tabulate as follows:

Bar A. Date:

Distance of centre of mass from heavier end of bar = 50·82 cms.

Distance from end to knife edge, AO cms.	$AG - AO = OG = h$ cms.	h^2	Time T	T^2	hT^2
1·60	49·22	2423	1·646	2·709	133·4
9·12	41·70	1739	1·591	2·531	105·5
19·50	31·32	981	1·546	2·390	74·8
30·71	20·11	404	1·597	2·550	51·3
35·32	15·50	240	1·688	2·849	44·2
	&c.		&c.		

With h^2 as abscissae and hT^2 as ordinates draw a curve for one half of the bar in the first and for the other half in the third quadrant (Fig. 28) and verify that it consists of two parallel straight lines which if produced intersect the h^2 axis at points equidistant from the zero.

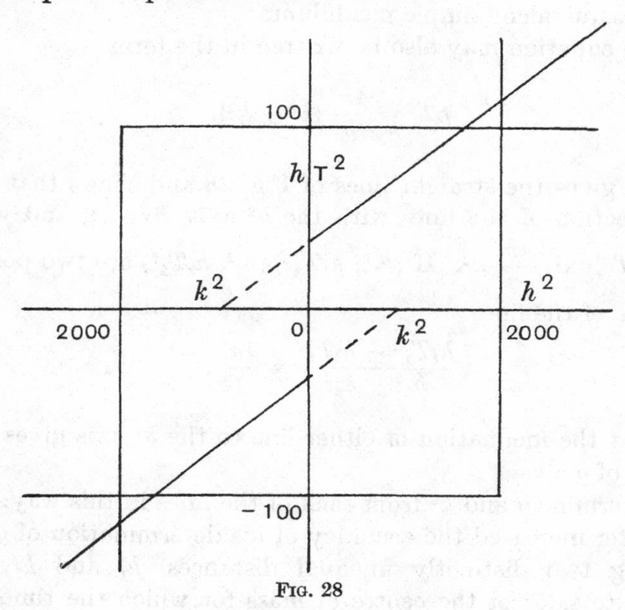

FIG. 28

For the compound pendulum we have

$$T = 2\pi\sqrt{\frac{k^2 + h^2}{gh}}\,*,$$

where k is the radius of gyration of the pendulum about an axis through its centre of mass parallel to the axis of rotation and g is the gravitational acceleration.

This equation may be written $T^2 = \frac{4\pi^2}{g}\left(\frac{k^2}{h} + h\right)$ which represents the curves of Fig. 26, which are therefore hyperbolae. They shew that there are in general four points along the axis of the bar, two on one side and two at equal distances on the other side of the centre of mass, about which the times of oscillation are identical. The distance

* See Lamb, *Dynamics*, chap. VIII, or Poynting and Thomson, *Properties of Matter*, chap. II.

from any one of these points to the non-symmetrical point on the other side of the centre of mass is the length $\frac{k^2}{h} + h$ of the equivalent simple pendulum.

The equation may also be written in the form

$$hT^2 = \frac{4\pi^2}{g}(k^2 + h^2),$$

which gives the straight lines of Fig. 28 and shews that the intersection of the lines with the h^2 axis give k^2, and with the hT^2 axis $\frac{4\pi^2}{g}k^2$. If $(h_1^2, h_1T_1^2)$ $(h_2^2, h_2T_2^2)$ are two points on one of the lines

$$\frac{h_1T_1^2 - h_2T_2^2}{h_1^2 - h_2^2} = \frac{4\pi^2}{g},$$

so that the inclination of either line to the h^2 axis gives the value of g.

Determine g and k^2 from each of the lines in this way.

Kater increased the accuracy of his determination of g by finding two distinctly unequal distances, h_1 and h_2, on opposite sides of the centre of mass for which the times of oscillation T_1 and T_2 were very nearly the same. Resolving the expression for $\frac{4\pi^2}{g}$ into partial fractions we have

$$\frac{T_1^2 + T_2^2}{h_1 + h_2} + \frac{T_1^2 - T_2^2}{h_1 - h_2} = \frac{8\pi^2}{g}.$$

The second term is very small and the value of g depends on $T_1^2 + T_2^2$ and on $h_1 + h_2$, the distance between the knife edges, which can be measured directly.

Take an observation obtained with the knife edge near one end, $h^2 > k^2$, combine with it one with nearly the same periodic time with the knife edge on the opposite side of G with $h^2 < k^2$ and calculate g. Similarly for the other end.

Measure directly the distances h_1 and h_2 in each case and use these measurements in calculating the value of the first term in the expression for $\frac{8\pi^2}{g}$. Slip the sliding weight on to the bar and adjust its position till the times T_1 and T_2 are as nearly as possible equal. Recalculate g.

In modern work the half-second pendulum has been most used. It is about 26 cms. long (Fig. 28 *a*) and can be enclosed in an evacuated vessel. The knife edge is fixed and the pendulum carries an agate plane at each end which rests on the knife edge.

The isochronism of the pendulum depends on the equality of $\sin\theta$ and θ for small values of θ. If θ becomes large the time of oscillation varies with the angle of displacement, and a closer approximation is given by

$$T = 2\pi\sqrt{\frac{l}{g}}\cdot\left(1 + \left(\frac{1}{2}\right)^2 \sin^2\frac{\theta}{2} + \left(\frac{1.3}{2.4}\right)^2 \sin^4\frac{\theta}{2} + \ldots\right),$$

where θ is the semi-arc of oscillation.

The following table gives the value of the factor in brackets for different semi-arcs of swing.

FIG. 28 *a*

θ	factor
1°	1·00002
5°	1·00048
10°	1·00191
20°	1·00766

SECTION XII

YOUNG'S MODULUS BY THE BENDING OF BEAMS

Apparatus required: *Uniform wooden beam and supports, slotted plate with hook and cross wire, weights, silvered glass scale and support.*

When a beam of breadth b and height h, supported at two points l apart but not clamped, carries a mass m at a point half-way between the supports, this point is depressed by an amount d given by the equation*

$$d = \tfrac{1}{4}m \, . \, g \frac{l^3}{\epsilon b h^3},$$

where ϵ is Young's modulus for the material of the beam, and g the gravitational acceleration.

To verify this relation, place the two supports provided, so that the top edges are in the same horizontal plane and about 80 cms. apart. Look across them to see that they are parallel.

Place the wooden beam with its greater breadth horizontal and the marks near its ends over the supports, and behind its middle point place a graduated mirror with the scale vertical (Fig. 29).

The load is applied by a light hook suspended from a small metal plate with a rectangular slot in it through which the beam passes. The plate has two horns projecting from its sides and between them a fine wire is stretched. Read the position of the wire on the scale, avoiding parallax.

By means of the hook suspend a 500 gram weight from the middle of the beam, and read on the scale the depression of the beam. Increase the weight to 1000 grams, and again read the depression. Verify, by means of your observations, that the depression is proportional to the deflecting weight. Turn the beam so that the top surface becomes the bottom one and repeat.

* See Lamb, *Statics*, chap. XVII. Poynting and Thomson, *Properties of Matter*, chap. VIII.

Turn the beam so that its greater breadth is vertical and repeat the observations, first with one edge downwards then the other.

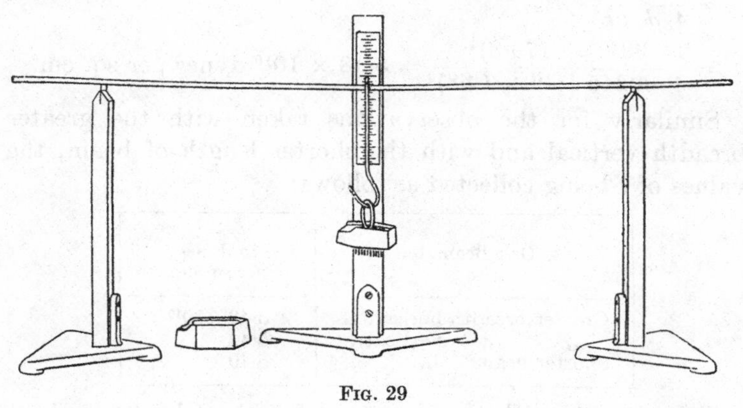

FIG. 29

Bring the supports nearer together so as to use a shorter length of the beam, and with the greater breadth horizontal repeat the observations.

From the observations calculate, using the above formula, Young's modulus for the material of the rod.

Arrange your results as follow:

Date:

Beam of Oak Wood marked O. 3.

Length of beam between supports ... = 79·8 cms.
Greater breadth ... ... = 1·36 cms.
Lesser breadth ... ... = ·88 cm.

Greater breadth horizontal.

Load in grams	Reading at centre	De-pression cms.	Reading at centre	De-pression cms.	Mean depression	Mean depression per gram added
0	8·10		7·90			
509	6·85	1·25	6·65	1·25	1·25	·00246
0	8·10		7·90			
963	5·80	2·30	5·50	2·40	2·34	·00244
0	8·10		7·90			
					Mean	·00245

The equality of the numbers in the last columns verifies the equation given. From it we have:

$$\epsilon = \frac{1}{4}\frac{mg}{d}\frac{l^3}{bh^3}$$

$$= \frac{1}{4}\frac{981}{\cdot00245}\frac{(79\cdot8)^3}{1\cdot36 \times (\cdot88)^3} = 5\cdot48 \times 10^{10} \text{ dynes per sq. cm.}$$

Similarly for the observations taken with the greater breadth vertical and with the shorter length of beam, the values of ϵ being collected as follows:

Oak Beam, 0·3	ϵ in dynes per sq. cm.
Greater breadth horizontal	$5\cdot48 \times 10^{10}$
" " vertical...	5·44
Shorter beam	5·49

Experiments with the shortened length of beam are here introduced in order to illustrate the law according to which the deflection is proportional to the cube of the length. If the sole purpose of the exercise had been to determine Young's modulus, these experiments would not have been necessary.

Reciprocal properties of points of application of load and measurement of deflection:

Place the beam in its first position with a hanger at the middle point and a similar one at a point which divides the beam in the ratio 1 : 3. Read the position of the cross-wire at the middle point, apply a weight of 500 grams at the other point and deflection at the middle point. Increase the weight to 1000 grams, and again read the deflection.

Load	Reading	Deflection cms.
0 at A	7·85 at B	
500 gr. at A	6·70 at B	1·15 at B
1000 gr. at A	5·65 at B	2·20 at B
0 at B	7·55 at A	
500 gr. at B	6·40 at A	1·15 at A
1000 gr. at B	5·40 at A	2·15 at A

Now place the mirror behind the point at which the weight has been applied, and remove the weight to the middle. Read the deflections for 500 and for 1000 grams, enter as below and verify that they are the same as those found in the former case. Hence the deflection at B due to a weight at A is equal to the deflection at A due to the same weight at B.

If metal bars are used the deflections will be smaller than in the case of wood, and must be read off by means of some form of reading microscope.

If the section of the beam though not rectangular is symmetrical about a horizontal line drawn through its centre we have

$$d = mg\,\frac{l^3}{48\epsilon I},$$

where I is the moment of inertia of the section about the horizontal through its centre.

Young's modulus may also be measured if the beam is fixed at one end and a weight is attached to the other end; the relation between Young's modulus and the deflection at the point of application of the force is in that case given by

$$d = mg\,\frac{l^3}{3\epsilon I} = 4m \,.\, g\,\frac{l^3}{\epsilon b h^3}.$$

If the beam has both its ends fixed so that they keep in line,

$$d = mg\,\frac{l^3}{192\epsilon I} = \frac{1}{16}\,mg\,\frac{l^3}{\epsilon b h^3}.$$

If the beam is inclined at θ to the horizontal g is replaced throughout by $g\cos\theta$.

SECTION XIII

MODULUS OF RIGIDITY

Apparatus required: *Uniform wires, cylindrical weight, and watch with seconds hand.*

The elastic reactions of a homogeneous body subject to strain depend on two constants, the bulk modulus, or resistance to change of volume, and the modulus of rigidity, or resistance to change of shape by simple shear.

If a cylinder AB of radius r and length l has one of its ends A fixed, while the other B is twisted about the axis of the cylinder, the angle of twist ϕ is connected with the moment of the twisting couple C, and the modulus of rigidity n by the relation

$$C = \frac{\pi}{2} nr^4 \frac{\phi}{l}\text{*}.$$

Hence if C and ϕ are measured, n can be calculated from the dimensions of the cylinder. C is generally applied by a known weight acting at a measured radius. The modulus of rigidity thus measured is sometimes called the "static rigidity." If the cylinder is thin, *e.g.* a wire, C may be determined by suspending a body of known moment of inertia from the wire, and letting it perform torsional oscillations. The modulus thus determined is sometimes called the "dynamic rigidity."

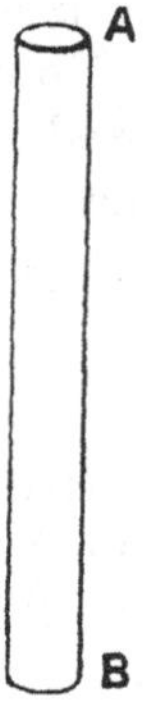

A

B

FIG. 30

The time T of a torsional oscillation will be

$$T = 2\pi \sqrt{\frac{I}{N}},$$

where I is the moment of inertia of the suspended body about

* See Lamb, *Statics*, chap. XVII. Poynting and Thomson, *Properties of Matter*, chap. VII.

the axis of twist, and N is the moment of the torsional couple per radian of twist, that is the value of C when $\phi = 1$, *i.e.*

$$N = \frac{\pi n r^4}{2l}.$$

Hence
$$T = 2\pi\sqrt{\frac{I}{\frac{\pi}{2}.n.\frac{r^4}{l}}}$$

and
$$n = \frac{8\pi I}{\frac{r^4}{l}.T^2}.$$

If the suspended body is of some regular form its moment of inertia about the torsional axis can be calculated from its mass and shape. Otherwise we may proceed as in the Exercise "Moments of Inertia II" p. 70, to determine I.

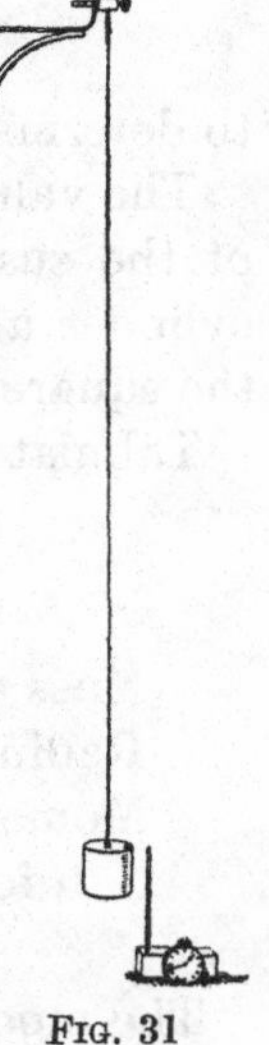

FIG. 31

In order to verify the truth of the law expressed by the above equation, suspend a brass cylinder (Fig. 31) by means of three different wires of the same material but different dimensions, and determine the times of torsional oscillation (Section III), p. 16. In the example given below two of the wires are cut from the same coil, the length of one piece being about double that of the other, and the third is equal in length to the longer of the two former wires, but is thicker. The diameters of the wires should be determined by the screw gauge at the ends and at several other points, dividing the wire into equal lengths.

If $r_1, r_2 \ldots r_n$, the radii at the ends and at points dividing the length between the ends equally, differ from each other by amounts greater than one part in 30, take as mean radius r where

$$\frac{1}{r^4} = \frac{1}{n-1}\left\{\frac{1}{2}\frac{1}{r_1{}^4} + \frac{1}{r_2{}^4} + \cdots + \frac{1}{2}\frac{1}{r_n{}^4}\right\},$$

and if they differ by less take the arithmetical mean.

Tabulate the results as follows:

Date: Steel wires marked C.

Length l cms.	Mean radius r cms.	r^4	$\frac{r^4}{l}$	T	T^2	$\frac{r^4}{l}.T^2$
83·5	·0165	$7{\cdot}41 \times 10^{-8}$	$8{\cdot}87 \times 10^{-10}$	6·45	41·60	$3{\cdot}69 \times 10^{-8}$
41·75	·0165	7·41 ,,	&c.	4·54	20·61	&c.
83·5	·0295	75·72 ,,	&c.	2·15	4·62	&c.
					Mean	

Take the mean of the values obtained in the last column and substitute in the equation

$$n = \frac{8\pi I}{\frac{r^4}{l}.T^2}$$

to determine the modulus of rigidity n.

The value of I can be found from the mass and dimensions of the suspended body, since the moment of inertia of a cylinder about its axis is the product of its mass into half the square of its radius.

Tabulate your results as follows:

Date:

Steel wires and cylinder marked C.

Mass of cylinder ... ... 663 grams

Radius of cylinder ... ... 2·27 cms.

Moment of Inertia (calculated) = 1578

$\therefore$ n for steel wire used = $8{\cdot}5 \times 10^{11}$ dynes per sq. cm.

The modulus of rigidity of the wire may also be found by applying to the cylinder it supports a couple of known moment C and measuring the resulting rotation ϕ of the cylinder as mentioned on p. 82. The couple is easily produced by two threads wrapped round the cylinder, which pass over pulleys and carry equal masses at their pendent ends. The rotation of the cylinder may either be measured directly by the help of a pointer attached to it and a circular scale below, or by the movement of a marked point on one of the threads.

SECTION XIV

VISCOSITY

Apparatus required: *Two long capillary tubes, an inverted bell-jar, a small flask, microscope and stage micrometer.*

If a liquid flows along a capillary tube the liquid in contact with the wall of the tube is either at rest or only moves very slowly. We shall assume that it is at rest. The layer of liquid next to the one in contact with the walls moves with small velocity, the layer next to it with a greater, and so on, the liquid at the axis of the tube moving with the greatest velocity. For the same tube, the velocity of the layers will increase more rapidly from the walls to the axis for a liquid of small viscosity than for a liquid of great viscosity. The volume of the liquid which under given conditions flows through the tube will therefore depend on the viscosity of the liquid.

It may be shewn* that the volume V is given by the equation

$$V = \frac{\pi p t}{8\eta l} . R^4, \quad = \frac{\pi p t}{8\eta l} \frac{2a^3b^3}{a^2 + b^2},$$

where

$R =$ radius, $l =$ length of the capillary tube.
a and $b =$ semi-axes of section if elliptic.
$p =$ excess of static pressure, *i.e.* when the flow is stopped, within the outlet of the tube over that without.
$\eta =$ coefficient of viscosity of the liquid.
$t =$ time in seconds.

The mean rel. of flow $v = \dfrac{p}{8\eta l} . R^2$.

* Lamb, *Hydrodynamics*, chap. XI. Poynting and Thomson, *Properties of Matter*, chap. XVIII.

An inverted bell-jar is provided (Fig. 32), having a rubber stopper through which passes a bent tube, to the end of which an inclined capillary tube of diameter $2R$ and length l is attached by a piece of rubber tubing provided with a clamp. A strip of paper is gummed to the jar about half-way up, and a horizontal mark on it serves to indicate the level of the liquid in the jar.

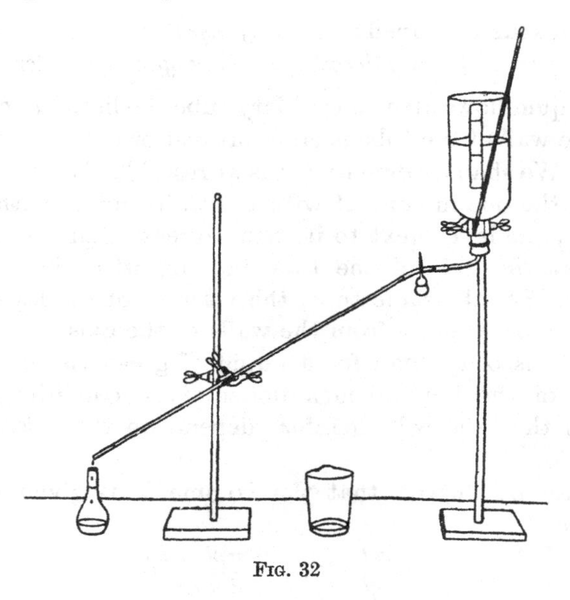

FIG. 32

Place the lower end of the capillary tube about 10 cms. above the table, and under it put a beaker to catch any liquid issuing. Place a thermometer in the bell-jar. Gum a piece of paper to the neck of a 4 oz. flask, and determine the volume up to the paper by weighing the flask when empty and when full of water. Fill the bell-jar with water at the temperature of the room, to a level a little above the upper mark on the gum paper. The water will flow through the tube into the beaker.

When the level of the water in the bell-jar has fallen to a convenient mark on the paper strip on the side of the jar, remove the beaker quickly and replace it by the flask, noting the time to a second. When the flask is filled to the level of the gum paper replace it by the beaker, notice again the time and mark on the paper strip the level of the water in the bell-jar.

Determine the excess of the mean static pressure p within the outlet of the tube over that without by measuring the heights of the two marks on the bell-jar above the centre of the bore of the tube at the outlet, and taking the mean, or by measuring the heights of the two marks from the table and subtracting the height of the centre of the bore of the tube above the table. If h is the mean height, ρ the density of the liquid, ρ' that of the fluid, generally air, surrounding the apparatus and g the gravitational acceleration,

$$p = hg(\rho - \rho').$$

The diameter of the tube should be found by placing the tube under the stage of a microscope having a scale in the eyepiece, or microscope provided with a cross wire and a graduated traversing screw, in such a way that one end is in focus and the axis of the tube is in line with that of the microscope. The greatest and least diameters of the bore of the tube at one end are thus found in divisions of the eyepiece scale. Reverse the tube and find the diameters at the other end. If the four measurements do not differ from each other by more than three per cent. take the mean or substitute

$$\frac{2a^3b^3}{a^2 + b^2} \text{ for } R^4,$$

if they differ materially use another tube.

To find the value of an eyepiece scale division in cms., remove the tube, place on the stage of the microscope a scale divided into millimetres and tenths, and find how many of the eyepiece divisions are equivalent to the millimetre of the stage scale.

Take a wider tube and determine by means of it the viscosity of the salt solution provided. Find the density of the solution by weighing the flask when filled with it up to the level of the paper.

Tabulate your results as follows:

Date:

Microscope A.

Divisions of eyepiece scale corresponding to 1 mm.	= 61·5.
∴ Value of a division	= ·00163 cm.
Weight of flask filled with water	= 100 grams.
Do. empty	= 24 grams.
∴ Volume of flask	= 76 c.c.
Weight of flask filled with solution	= 116 grams.
∴ Density of solution	= 1·21.

Viscosity of water. Tube A.

Diameters (1) 68·6, 68·4; (2) 67·7, 67·3	= 68 divisions of eyepiece scale.
	= ·111 cm.
Length	= 65·5 cms.
Volume of water	= 76 c.c.
Time	= 717 seconds.
Pressure	= 19·6 cms. water.
	= 19230 dynes per sq. cm.
Temperature	= 11°·5 C.
η at 11°·5 C.	= ·0103.

And similarly for the observations with the solution.

Take a tube of about ·3 cm. bore and repeat the experiment with water, making p in the first instance small and determining the flow per second. Increase p by steps, noting the flow at each step. Notice that the flow is proportional to the pressure up to a point at which the water in the tube begins to have eddies in it and the rate of increase of flow per unit increase of pressure diminishes. Plot a curve with flow as ordinates and p as abscissae and shew that eddies begin when $\frac{vR}{\eta/\rho} = 1000$ about.

When a gas is substituted for a liquid the same expression holds for the flow, but p is now the excess of pressure at the entry over that at the exit of the capillary tube and V is the volume the gas delivered would have if measured at the mean of the pressures at entry and exit.

SECTION XV

SURFACE TENSION

Apparatus required: *Glass tube and scale, microscope and stage micrometer.*

Heat in the blow-pipe flame a piece of glass tubing which has been *thoroughly cleaned inside* with water and dried, and when a portion about a centimetre long, four or five centimetres from one end, is soft draw the two ends apart so as to form a fine capillary tube.

Break off the capillary part of the tube, and repeat the drawing-out process on the rest of the tube by heating a portion close to that previously heated, using the contracted drawn-out end to hold by, till half-a-dozen capillary tubes have been obtained, two about 1, two about ·7, and two about ·3 mm. diameter. The drawing-out process should be carried out more slowly the wider the capillary tube required.

Break off from each of the finer tubes about 20 cms. and from the wider about 10 cms. of the most uniform portion.

Mount vertically in a small glass vessel containing water a transparent graduated scale. Both vessel and scale should be thoroughly clean. Wet the scale a little and place a tube against it (Fig. 33); the tube will stick and will not need supporting. The water will ascend in the tube. When the water has reached its greatest height, raise the tube a few mms., so that the inner wall above the meniscus is wet, and read on the scale the

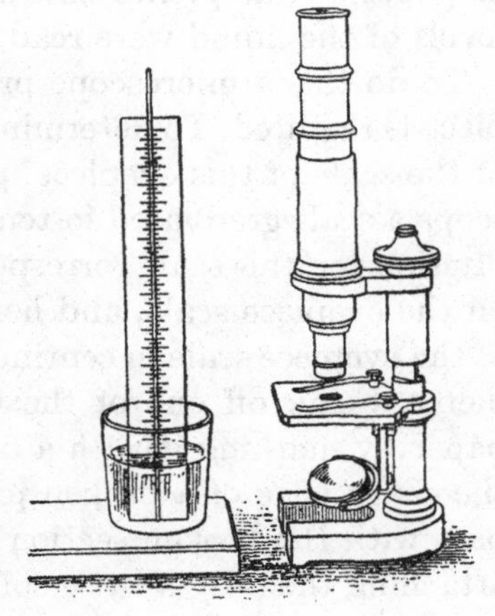

Fig. 33

levels of the meniscus and of the surface of the liquid in the vessel.

In taking the latter reading the level of the surface at a little distance from the scale should be read, since close to the scale the surface is curved upwards, thus: Read A, not B.

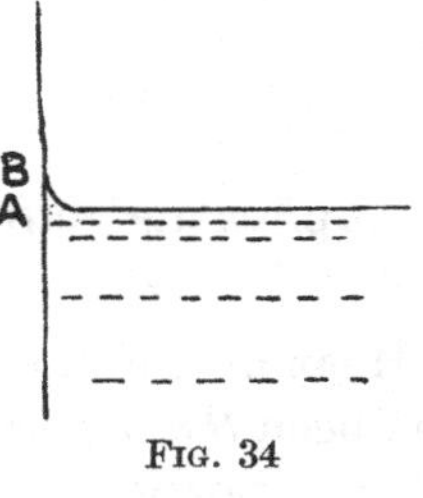

Fig. 34

Gum a small piece of paper to the tube, 2 mms. above the top of the column of liquid in it.

Raise the tube 5 mms., and again take readings. If the difference of level of the meniscus inside the tube and of the liquid outside is the same as it was previously, the tube is uniform. If there is much difference another tube should be substituted for it.

Repeat the observations with two other tubes of different diameters, attaching a strip of paper to each.

Now take another vessel, clean it thoroughly, fill it with 20 % ethyl alcohol, and take observations with the other three tubes, attaching strips of paper as before. Pour the alcohol back into the stock bottle.

The radii of the tubes at points 4·5 mms. below the strips of paper, *i.e.* at points half-way between those at which the levels of the liquid were read, must now be found.

To do this a microscope provided with a micrometer eyepiece is required. To determine in cms. the value of a division of the scale of this eyepiece, place on the stage of the microscope a scale graduated in tenths of a mm., notice how many divisions on this scale correspond to some convenient number on the eyepiece scale, and hence find the value of a division of the eyepiece scale in centimetres. Remove the stage micrometer, break off one of the tubes 45 mms. below the gum paper by marking it with a fine file or rough-edged knife, or the sharp edge of a broken piece of quartz, and mount each part with the broken section upwards on a microscope slide attaching them by a little soft wax. Now place one of these mounted tubes on the stage, focus the broken end of the tube, and read its diameter in scale divisions of the eyepiece micrometer. If it is not accurately circular, measure its greatest and least diameters and take the mean.

Repeat with the other piece. Take the mean and convert to cms.

Do this with each capillary tube used, and calculate the value of the surface tension T in dynes per cm. from the equation

$$T = \frac{g\rho hr}{2}\left(1 + \frac{1}{3}\frac{r}{h}\right),$$

where

g = value of gravitational acceleration.
ρ = density of liquid.
h = height of meniscus in tube above liquid outside.
r = radius of tube*.

Tabulate your results as follows:

Date:

Microscope A.

10 divisions of eyepiece scale = 1·67 divisions on stage scale = ·167 mm.

∴ 1 division of eyepiece scale = ·00167 cm.

Liquid	Tube	Reading in cms. at		Height cms.	Radius		$h.r.$	T
		Meniscus	Base		eyepiece divisions	cms.		
Water	1	3·66	1·52	2·14	42·5	·071	·152	74·2
		3·65	1·51	2·14				
	2	6·65	1·53	5·12				
		6·64	1·52	5·12				
	3							
Alcohol 20 %	4							
	5							
	6							

* See Poynting and Thomson, *Properties of Matter*, chap. XIV. If the tube is elliptical with semi-axes a and b, r may be taken as

$$\frac{2ab}{\frac{a+b}{2} + \sqrt{\frac{a^2+b^2}{2}}}.$$

Verify the value found for water by the following methods:

Balance Method.

Clean carefully the thin sheet of glass provided, and suspend it from the left-hand end of the beam of a simple balance by means of the thin metal clip. Adjust the sheet till its under edge is horizontal, and place weights in the right-hand pan till the pointer of the balance reads zero. Fill a clean glass vessel, wider than the strip of glass, with clean water, place the vessel under the strip and raise it gradually till the lower edge of the glass just touches the surface of the water. Push the glass a millimetre down into the water so as to wet the edge, and then place weights in the right-hand pan of the balance till the edge is just torn away from the water. If m grams, in addition to the weight to balance the glass in air, have to be used, and if l is the length of the edge of glass touching the surface,

$$mg = T \,.\, 2l, \text{ or } T = \frac{mg}{2l}.$$

Record as follows:

Weight to balance glass in air	=	·52 gram
,, tear glass from water	=	1·59 grams
Difference due to surface tension	=	1·07 ,,
Length of edge touching water	=	7·5 cms.
Surface tension of water	=	70 dynes per cm.

Falling Drop Method.

Connect about 4 cms. of glass capillary tube, about ·5 mm. internal and 6 mms. external diameter, by means of 20 or 30 cms. of rubber tube to the outlet of a small funnel. The three should be clean and the outlet end of the capillary tube regular and perpendicular to the axis. Place a beaker under the outlet and fill the funnel with water. Raise the capillary tube till its outlet is nearly on the same level as the surface in the funnel and one drop falls from the tube per minute about.

Weigh a small beaker and in the interval between the fall of two drops substitute it for the first beaker. Allow 10 drops to fall, substitute the first beaker and weigh the second. Insert it again to receive another 10, weigh again and repeat till 50 drops have been collected. Subtract the original weight from that with 30, that with 10 from that with 40, &c., and take the mean of the differences as the mass of 30 drops. Let m be the mass of one drop and r the outer radius of the end of the capillary tube, then

$$T = \frac{mg}{3{\cdot}8r}*.$$

Measure the outer radius of the capillary tube by means of a screw gauge, taking the mean radius if it is not quite uniform.

Record as follows:

Weights of beaker.

Drops	Weight	Drops	Weight	Difference
0	70·102	30	75·105	5·003
10	71·773	40	76·778	5·005
20	73·446	50	78·452	5·006
			Mean	5·005

$\therefore$ m = ·168 gram.

Radius: ·598, ·603, mean ·601.

T = 72·1.

* Rayleigh, *Papers*, IV, p. 415.

BOOK III

HEAT

SECTION XVI

COEFFICIENT OF EXPANSION OF A SOLID

Apparatus required: *Tubes of metal and glass, flow of hot water or steam, thermometer and reading microscope.*

When the temperature of a solid is raised, the solid generally expands, and if the distance between two given points in the solid is l_0 at 0° C. and l_t at t° C. we have, to a sufficient degree of approximation,

$$l_t = l_0 (1 + at),$$

where a is a constant called the coefficient of linear expansion.

In order to determine this coefficient, a bar of convenient length may be measured at two temperatures t_1 and t_2. If l_1 and l_2 are the observed lengths, we have as a consequence of the above relation

$$a = \frac{l_2 - l_1}{l_0 (t_2 - t_1)}.$$

As a is small in the case of solids, we may as a rule with sufficient accuracy substitute the length at the ordinary temperature of the room for l_0. Thus the coefficient of expansion of metals generally lies between ·00001 and ·00004, and their linear dimensions will increase therefore by less than a thousandth part of the original length between 0° and 20°. If therefore it is not desired to measure a to an accuracy greater than one part in a thousand, l_0 may be taken as identical with the length measured at the ordinary temperature of the room, and with rods about half a metre long an ordinary scale graduated to millimetres will be sufficient for the purpose of measurement.

The quantity $l_2 - l_1$ being small some method of measurement suitable for small lengths must be used, *e.g.* a microscope provided with a graduated scale in the eyepiece.

The heating of the material to be tested is carried out very conveniently if it is given in the form of a hollow rod or tube, since hot water or steam may be sent through it to raise it to the required temperature. If it is a solid rod it should be supported in the axis of a glass tube through which the steam or water is sent.

If two reading microscopes are available, two marks may be made on the tube or rod, one near each end, and the displacements of these marks, after a temperature change which should not be less than 70° C., measured by means of the microscopes, which should be clamped to a solid base. The apparatus provided gives sufficiently accurate results with the mark at one end of the tube fixed and the other observed through a reading microscope only.

Two tubes of about 60 cms. length are provided, one of glass and one of brass. They can be mounted so that steam or hot water may be passed through them.

A notch is cut near one end of each tube, and this notch with the tube placed in a horizontal position is made to rest on the knife edge of the stand provided. For additional security the tube may be clamped above the knife edge. The other end of the tube may rest on the microscope stage, but in the stand provided it is supported by passing loosely through a hole a piece of wood at one end of a paper protecting tube fixed to the stand.

The free end of the tube has a small patch of white paint on it across which a fine scratch or a fine pencil line is drawn at right angles to the tube. A microscope is provided, having an objective of low power capable of including a field of view of about 2 mm. diameter, and an eyepiece with a micrometer scale. Place the microscope in such a position that the pencil

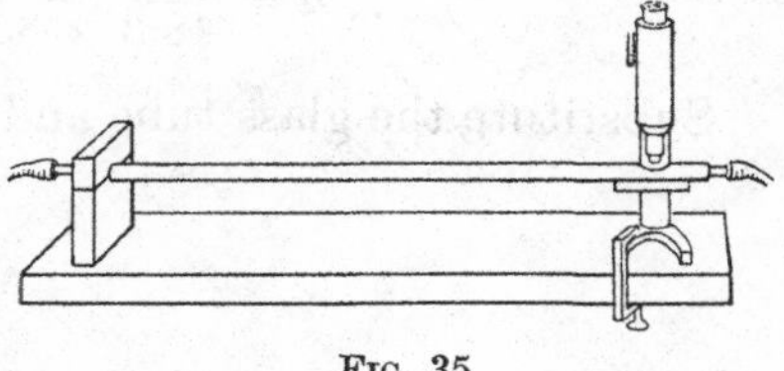

FIG. 35

line is near the centre of the field and parallel to the divisions on the scale and then clamp the microscope firmly to the table or slab (Fig. 35).

Attach rubber tubes to the ends of the experimental tube and send a stream of cold water through; observing the temperature of the water by means of a thermometer placed in the issuing stream. Read the position of the line in the microscope.

Send steam or hot water through the tube and again take readings. Finally repeat the experiment with cold water.

Measure the length between the notch and the pencil line on the tube to the nearest mm.

Determine the value of a scale division of the eyepiece by placing a scale graduated in $\frac{1}{10}$ mm. on the stage of the microscope.

Record and calculate the coefficient as follows:

Date:

Microscope *A*.

58 divisions of eyepiece scale = 10 on stage scale = 1 mm.

$\therefore$ 1 division of the eyepiece = ·00172 cm.

Brass tube 48·3 cms. between marks.

Temp.	Reading	Difference	
		Scale	Cms.
Water 15° C.	1·42		
Steam 100	5·72	4·30	·0744
Water 15	1·42		

$$\therefore a = \frac{\cdot 0744}{48\cdot 3 \times 85} = \cdot 0000181.$$

Substitute the glass tube and record similarly.

SECTION XVII

THERMAL EXPANSION OF A LIQUID

Apparatus required: *Bulb with graduated stem, and ice.*

The change of volume of a liquid with increase of temperature is most conveniently found by enclosing the liquid in a vessel made of a material of which the coefficient of expansion is known, and observing the apparent change of volume of the liquid as the temperature of liquid and containing vessel is varied.

Since the apparent expansion is, in general, only a small fraction of the total volume, the volume of that part of the vessel in which the expansion is measured should only be a small fraction of the total volume.

The vessel, which is called a "dilatometer," is generally a glass bulb provided with a graduated stem having a fine bore, the liquid filling the bulb and part of the stem (Fig. 36).

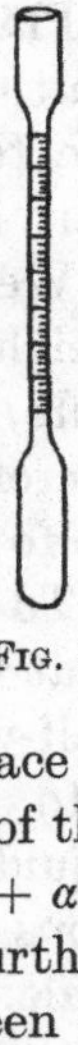

If at 0° C. the surface of the liquid in the stem stands at the division n_0 of the stem, and the volume of the bulb up to the zero division of the scale is m, measured in scale divisions, the volume of the liquid at 0° C. is $m + n_0$ scale divisions.

Fig. 36

If when the dilatometer and liquid are at $t°$ C. the surface of the liquid reads n on the scale, we have, for the volume of the vessel up to the mark n at temperature $t°$ C., $(m + n)(1 + at)$, where a is the coefficient of expansion of the vessel. If further β_t is the expansion of unit volume of the liquid between 0° and $t°$, unit volume at the freezing point of water becomes $1 + \beta_t$ at temperature t, and the volume of the liquid at the latter temperature will be $(m + n_0)(1 + \beta_t)$, hence:

$$(m + n)(1 + at) = (m + n_0)(1 + \beta_t)$$

$$\text{or } 1 + \beta_t = \frac{m + n}{m + n_0}(1 + \alpha t)$$

$$= 1 + \frac{n - n_0}{m} + \alpha t \qquad \text{approximately.}$$

$$\therefore \beta_t = \frac{n - n_0}{m} + \alpha t,$$

from which the expansion β_t may readily be determined.

A graduated capillary tube is provided, at the end of which a bulb of about 1 cm. diameter has been blown.

Wash and dry the bulb and tube and then weigh.

Heat the bulb gently in a Bunsen burner, then dip the open end of the tube into a small quantity of the liquid to be tested which, if necessary, has been boiled to free it from dissolved air. As the air in the bulb contracts, liquid is drawn up the tube into the bulb. Repeat the heating and cooling till the bulb and a short length of the stem are filled with liquid.

Place the tube in a water bath at about 16° C., and after waiting 10 minutes to allow the liquid in the bulb to take up the temperature of the bath, read the temperature of the bath and the position of the liquid in the stem.

Weigh the bulb and contents. The difference between this weight and the previous one is the weight of the liquid in the bulb, and if its density is known, the volume of the liquid, therefore the volume of the bulb and stem containing it, can be found.

To determine the volume of one scale division of the stem, heat the bulb gently till the liquid stands rather more than half-way up the stem, then plunge the open end of the stem into clean mercury, which will be drawn up the tube as the liquid contracts. Allow the dilatometer to cool till the positions of both ends of the mercury thread can be read on the scale.

Observe the length of the thread and weigh the bulb and its contents again. The difference between this weight and the last one is the weight of mercury, which may be divided by 13·596 to give the volume, and hence the volume of a scale division of the stem in c.c. may be found.

The dilatometer should again be heated till the mercury thread is expelled from the tube, and the upper end of the tube then sealed in the blow-pipe (Fig. 36). A label is attached to the tube stating the liquid in the tube, the volume of the bulb, and of a scale division of the stem; from these numbers the quantity m may be calculated.

The dilatometer should first be placed in a bath containing melting ice, and after the position of the liquid has become constant, a reading should be taken. The temperature of the bath should then be raised about two degrees, and observation of temperature and apparent volume taken at the new temperature. The observation should be repeated at intervals of 2° C. up to 10° C. and then at intervals of 5° C. over the range of temperature desired, and again as the temperature is decreased.

Care must be taken to wait sufficiently long after each change of temperature to allow the liquid in the bulb to come to its final state, for it must be remembered that the thermal conductivity of liquids is small, and that convection currents are diminished owing to the small size of the bulb.

The observation and results should be tabulated as follows:

Expansion of Water.

Date: Dilatometer A, $m = 15500$, $\alpha = \cdot 000020$.

Temp.	Reading	$n - n_0$	$\frac{n - n_0}{m}$	at	β_t
0° C.	24·0	0	0	0	0
0·6	23·2	− ·8	− ·0000516	·000012	− ·000040
1·7	22·1	− 1·9	− ·000119	·000034	− ·000085
3·3	21·1	− 2·9	− ·000192	·000066	− ·000126
5·3	20·6	− 3·4	− ·000222	·000106	− ·000116
7·3	21·4	− 2·6	− ·000176	·000146	− ·000030
12·3	25·5	+ 1·5	+ ·000098	·000246	+ ·000344
&c.	&c.		&c.		

Numerical calculation will be facilitated if the reciprocal of m is first calculated and then multiplied successively by the numbers in the third column. Three figures are sufficient and Crelle's Tables or a slide rule may be used.

A curve (Fig. 37) should be drawn with the observed temperatures as abscissae and total volumes $1 + \beta_t$ as ordinates.

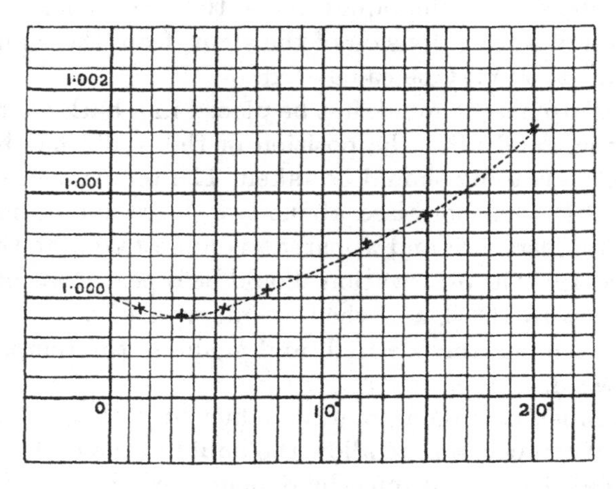

FIG. 37

The numbers taken with rising and falling thermometer should be plotted separately and should shew no systematic difference.

If the liquid is water especial care must be taken with the observations between 0° and 10° owing to the anomalous behaviour of that liquid.

Numbers giving the volume of 1 gram of water at different temperatures should be taken from Kaye and Laby's *Physical Tables*, and represented by a second curve for comparison.

SECTION XVIII

COEFFICIENT OF INCREASE OF PRESSURE OF A GAS WITH TEMPERATURE

Apparatus required: *Bulb and pressure apparatus, water bath, thermometer*.

If the temperature of a constant volume of gas is raised, the pressure of the gas varies according to the equation $p_t = p_0 (1 + \beta t)$, where p_0 is the pressure at 0° C., t the temperature Centigrade, and β a constant, called the coefficient of increase of pressure with temperature.

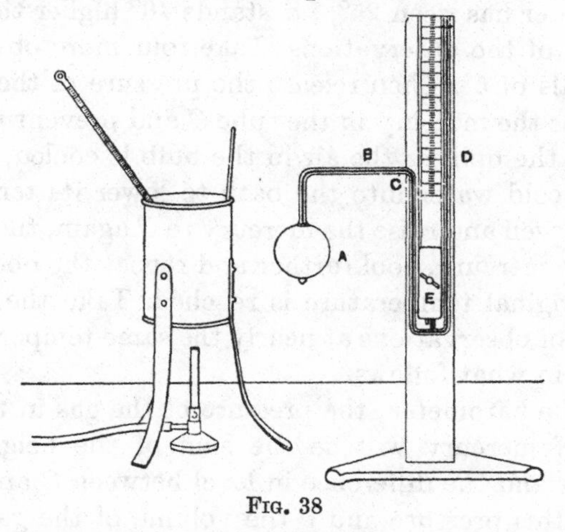

Fig. 38

To verify the law expressed by the above equation in the case of air the apparatus shewn in Fig. 38 is provided.

A is a glass bulb to which a bent tube of small bore BCD is attached. At the lowest point of the tube it branches and to the branch is attached a piece of rubber tubing which can be compressed between the plate E and the stand by turning a screw.

The rubber tube and the lower part of the tube CD are

filled with mercury, and the open end of the rubber tube then closed by a short glass rod.

By adjusting the quantity of mercury and the pressure at E the surface of the mercury in the left-hand tube may be brought to the fixed mark C. If the position of the surface of the mercury in D is then read, and the height of the barometer is known, we have the total pressure to which the air in A is subjected.

To determine how this pressure varies with the temperature of the air in the bulb, fill the bath in which the bulb is placed with cold water, noting its temperature. Adjust the mercury to C, and read the position of D on the scale alongside.

Raise the temperature of the bath about 5° C., keeping the water stirred, and repeat the adjustment and readings. Take four observations at intervals of 5° C. and then wait until the thermometer has risen 25°, *i.e.* stands 40° higher than at the beginning of the observations. Take four more observations at intervals of 5°. Then release the pressure of the screw so as to lower the mercury in the tube C and prevent it running back into the bulb as the air in the bulb is cooled, and pour sufficient cold water into the bath to lower its temperature 5° C. Stir well and raise the mercury to C again, then read D. Lower the mercury, cool further and repeat the observations till the original temperature is reached. Take the means of each pair of observations at nearly the same temperature and use them in what follows.

Read the barometer; the pressure of the gas in terms of a column of mercury will be the sum of the height of the barometer and the difference in level between C and D.

If p is this pressure and v the volume of the gas at temperature t, we have $pv = p_0 v_0 (1 + \beta t)$, where v_0 is the volume at 0° C.

The volume of the containing vessel will increase slightly as the temperature rises owing to the expansion of the glass.

If the coefficient of expansion of the glass is α, the volume v of the vessel is obtained from $v = v_0 (1 + \alpha t)$.

Hence $$p = p_0 \frac{1 + \beta t}{1 + \alpha t}.$$

Or since in the above experiments at is always less than ·002 and βt less than ·2, we may write with an accessory of 1 part in 2000

$$p = p_0 (1 + (\beta - \alpha) t).$$

If $p_1 t_1$, $p_2 t_2$ are two sets of readings of p and t, we may form two similar equations and eliminating p_0 find

$$\frac{p_2}{(1 + (\beta - \alpha) t_2)} = \frac{p_1}{1 + (\beta - \alpha) t_1},$$

and hence

$$\beta - \alpha = \frac{p_2 - p_1}{p_1 t_2 - p_2 t_1} = \frac{\frac{1}{p_1} - \frac{1}{p_2}}{\frac{t_2}{p_2} - \frac{t_1}{p_1}};^{*}$$

β may therefore be calculated, if α is known, by observation of any two corresponding pressures and temperatures. If a number of readings are taken at regular intervals of temperature we may apply the method explained in Section III and secure the smallest probable error by taking a certain number of observations, omitting the same number, and finally taking again an equal number. The observations are combined and reduced as given in the table below. Of the two forms for $(\beta - \alpha)$ given above the first involves less arithmetical labour, and would naturally be used in case the pressure is observed for only two temperatures. But for a series of experiments the results may be tabulated in a more symmetrical manner if the second form is adopted. The labour of calculation is reduced if the reciprocals of p_1 and p_2 are obtained from Barlow's Tables.

The fractions t_2/p_2 &c. and the final quotients may be found with sufficient accuracy from Crelle's multiplication tables.

* For more accurate work the expression,

$$\beta = \frac{\frac{1}{(1 + at_1) p_1} - \frac{1}{(1 + at_2) p_2}}{\frac{t_2}{(1 + at_2) p_2} - \frac{t_1}{(1 + at_1) p_1}}$$

should be used.

Arrange the observations as follows:

Date:

Apparatus marked K. Height of Barometer 76·2 cms.

t_1	p_1	$\frac{1}{p_1}$	$\frac{t_1}{p_1}$	t_2	p_2	$\frac{1}{p_2}$	$\frac{t_2}{p_2}$	$\frac{1}{p_1}-\frac{1}{p_2}$	$\frac{t_2}{p_2}-\frac{t_1}{p_1}$	$\beta-\alpha$
10°·5	77·7	·01287	·135	50·3	88·4	·01131	·567	·00156	·432	·00361
15·2	78·9	1267	·191	55·1	89·6	1116	·609	151	·418	361
20	80·1	1248	·249	60·0	90·9	1100	·660	148	·411	360
25·0	81·3	1230	·307	65·0	92·1	1086	·706	144	·399	361

Mean value of $(\beta-\alpha)$ ... ... $=$ ·003608

For glass of bulb α ... ... $=$ ·000028

β for air ... ... $=$ ·003636

$-1/\beta =$ absolute zero of constant volume air thermometer ... ... $= -275°$ C.

According to Regnault β should be equal to ·003669. There are two further sources of error which tend to reduce the apparent value of β. One is the expansion of the glass vessel due to increased internal pressure at the higher temperatures, and the other is due to the air in the tube leading from the bulb not being at the same temperature as that in the bulb. The first-named error is not likely to be more than 1 in 300,000 with the dimension of apparatus used, the second error will not exceed 1 in 6000 if the bulb has a diameter of 10 cms., and the length of capillary tube between the bulb and the fixed mark is not more than 20 cms. and its diameter not more than 18 mms.

If the interior of bulb is not thoroughly dry too high value will be obtained owing to the evaporation of the water from the walls of the vessel. To secure dryness a few drops of strong sulphuric acid are introduced into the bulb before the mercury is poured into the tube.

The increase of pressure of a gas is used to measure temperatures in the "constant volume gas thermometers," the construction of which is the same in principle as that of the apparatus used in this exercise.

SECTION XIX

COEFFICIENT OF EXPANSION OF A GAS AT CONSTANT PRESSURE

Apparatus required: *Graduated tube and index, water bath, and thermometer.*

When the temperature of a gas maintained at constant pressure is raised, the volume increases, and may be represented by an expression of the form

$$v = v_0 (1 + \alpha t),$$

where v is the volume at $t°$ C., v_0 at 0° C. and α is a constant called the coefficient of expansion of the gas.

If the gas is enclosed in a vessel the coefficient of cubical expansion of the material of which is β, the *apparent* volume of the gas will for expansions not greater than 30 per cent. be expressed with an accuracy of 1 part in 2000 by the equation

$$v = v_0 (1 + \overline{\alpha - \beta}\, t).$$

If t_1 and t_2 are two temperatures and v_1, v_2 the corresponding apparent volumes, it is found on eliminating v_0 that

$$\alpha - \beta = \frac{v_2 - v_1}{v_1 t_2 - v_2 t_1}$$

$$= \frac{\dfrac{1}{v_1} - \dfrac{1}{v_2}}{\dfrac{t_2}{v_2} - \dfrac{t_1}{v_1}}.$$

Hence if observations of the variation of volume of a given mass of gas, enclosed in a vessel of known coefficient of expansion, are made, the coefficient of expansion of the gas may be determined.

The apparatus provided consists of a graduated glass capillary tube, one end of which is sealed, and the other attached to a wide tube bent at right angles with its open end constricted. The complete tube is attached to a brass frame. The capillary tube is filled for about two-thirds of its length with air, which is confined by a thread of strong sulphuric acid in which a few grains of indigo have been dissolved. This thread serves both to indicate the volume of the air and to keep it dry (Fig. 39). The wider part of the tube serves as a reservoir of acid whose height is not appreciably affected by the flow of acid into or out of the capillary tube. The pressure to which the enclosed air is subjected remains therefore constant whatever its volume.

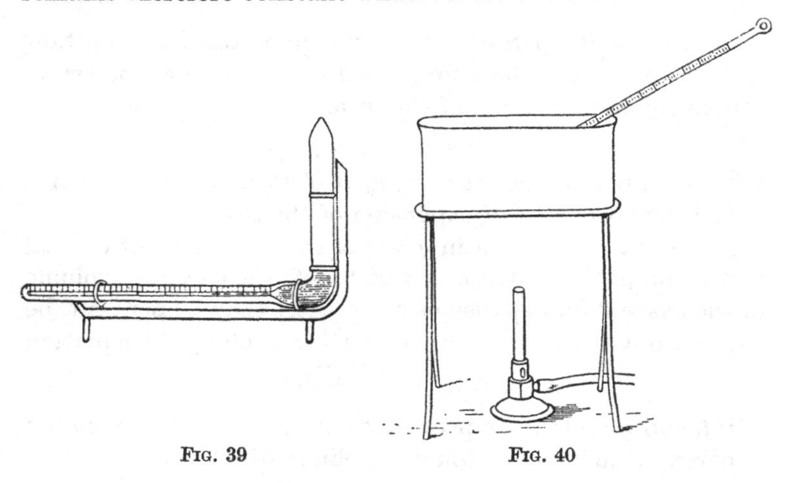

Fig. 39 Fig. 40

To find the effect of change of temperature on the enclosed air, place the graduated tube with the closed end about ·5 cm. higher than the acid index in a water bath the temperature of which can be varied (Fig. 40). Mix ice and if necessary salt with the water till the temperature indicated on a thermometer placed in the water is 0° C. Stir the water well, and after waiting till the thread of acid has taken a constant position, take a reading of that end of it which is in contact with the gas. Increase the temperature of the water by 5° C., stir it well and again observe the volume when it has become constant. Take four observations, increasing the temperature in steps of 5° C., then wait until the temperature has risen another 25° and take four more observations at temperatures increasing by 5° C.

Then decrease the temperature and take observations at approximately the same temperatures as during the rise.

The closed end will not in general be at the zero of the scale, and its reading must therefore be subtracted from the readings of the index to obtain the volumes. As the closed end is often conical a short thread of acid may be introduced to occupy the conical space. It is assumed that the bore is uniform so that the differences in the readings of the scale are proportional to the volumes.

The observations are reduced as in the previous section.

Enter as follows

Apparatus *B*. Date:

Reading at closed end 5·05.

Temperature increasing			Temperature decreasing		
t	Reading	Vol.	t	Reading	Vol.
0·0	15·25	10·20	·0	15·25	10·20
5·0	15·42	10·37	4·8	15·42	10·37
10·1	15·59	10·54	10·2	15·61	10·56
15·1	15·80	10·75	14·1	15·76	10·70
39·5	16·67	11·62	40·5	16·71	11·66
43·8	16·83	11·78	44·2	16·87	11·82
49·8	17·03	11·98	50·6	17·07	12·02
54·6	17·23	12·18	55·6	17·27	12·22

Take the mean of the numbers for increasing and decreasing temperatures and enter as in the following table.

t_1	v_1	$\frac{1}{v_1}$	$\frac{t_1}{v_1}$	t_2	v_2	$\frac{1}{v_2}$	$\frac{t_2}{v_2}$	$\frac{1}{v_1}-\frac{1}{v_2}$	$\frac{t_2}{v_2}-\frac{t_1}{v_1}$	$\alpha-\beta$
0°C.	10·20	·0980	·0	40°·0	11·64	·0859	3·44	·0121	3·44	·00353
4·9	10·37	·0964	·47	44·0	11·80	·0847	3·73	·0117	3·26	·00359
10·15	10·55	·0948	·96	50·2	12·00	·0833	4·18	·0115	3·22	·00357
14·6	10·72	·0933	1·36	55·1	12·20	·0820	4·52	·0113	3·16	·00358

Mean value of $\alpha - \beta$ ... ... $=$ ·00357

β for glass used ... ... $=$ ·00003

$\therefore \alpha$ ... ... $=$ ·00360

$-1/\alpha =$ zero of constant pressure air thermometer ... ... $= -$ 277° C.

For gases satisfying Boyle's law the coefficient of expansion at constant pressure is the same as that of increase of pressure at constant volume. The value of α found is about two per cent. too small, which may be accounted for by the capillary tube being wider near the division 15 than at lower readings.

Take temperatures as abscissae, and volumes in scale divisions as ordinates and draw a straight line passing as nearly as possible through the points so determined (Fig. 41). From the inclination of the straight line find the value of $\alpha - \beta$, and compare it with the result of the calculation.

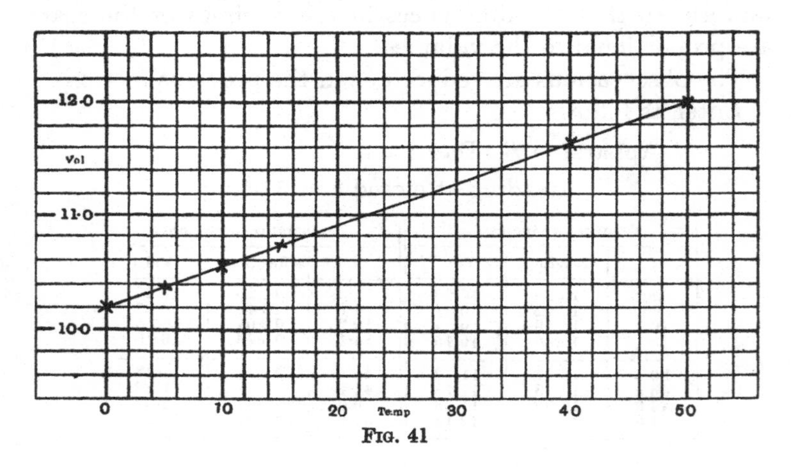

Fig. 41

SECTION XX

EFFECT OF PRESSURE ON THE BOILING POINT OF A LIQUID

Apparatus required: *Thermometer, flask, condenser and filter-pump.*

When the pressure under which a liquid boils is increased, the temperature to which the liquid must be brought before boiling takes place is raised, and if the pressure is decreased the temperature at which boiling occurs is reduced. It is the object of the present exercise to find the relation between the two changes.

In the diagram of the apparatus to be used (Fig. 42), A is a Bunsen filter pump through which a rapid flow of water from the tap T_1 may be kept up.

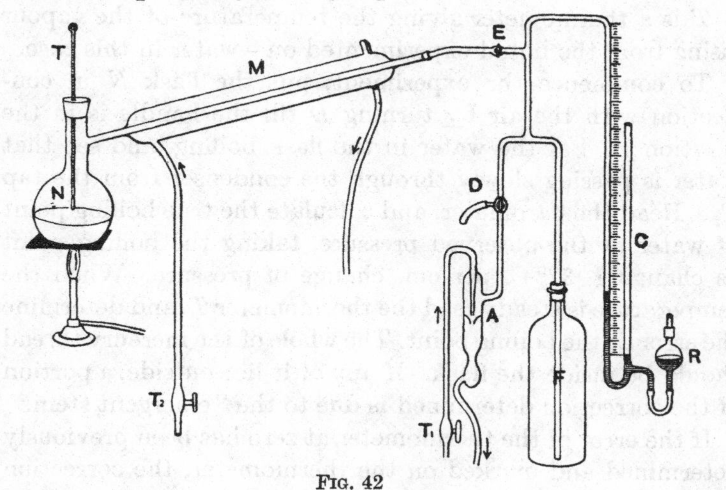

Fig. 42

D is a three-way cock having a central passage parallel to the handle which is joined at the centre by another passage at right angles leading only to one side. The side to which it leads is marked by a black dot. When the handle is vertical with the black dot to the right, the pump is in connection with the condensing apparatus. Call this position A. When the black dot is to the left the apparatus is open to the air. Call this position B. When the handle is horizontal with the black dot below, the way to the bottle is closed, and the pump is in connection with the air. Call this position C.

F is a bottle to diminish the changes of pressure due to the bursting of bubbles in the flask *N* during boiling, so arranged as to prevent water from the pump flowing back to the manometer when the pump is stopped.

E is a stop-cock to cut off the boiling apparatus from the pump.

C is a mercury manometer, with a reservoir *R* for adjusting the reading of the surface in the open tube to the zero of the scale alongside.

M is a condenser in which the steam is condensed, the condensed liquid running back into the flask.

N is the flask in which the boiling takes place.

T is a thermometer giving the temperature of the vapour rising from the liquid experimented on—water in this case.

To commence the experiment, put the flask *N* in connection with the air by turning *D* till the handle is in the position B. Set the water in the flask boiling, and see that water is passing slowly through the condenser from the tap T_2. Read the barometer, and calculate the true boiling point of water at the observed pressure, taking the boiling point as changing ·37° C. per cm. change of pressure. When the temperature is steady read the thermometer *T*, and determine the error at the boiling point. The whole of the mercury thread should be inside the flask. If any of it lies outside, a portion of the correction determined is due to the "emergent stem."

If the error of the thermometer at zero has been previously determined and marked on the thermometer, the correction at any reading may be found with sufficient accuracy for the present purpose* by drawing the curve of corrections as in the *Intermediate Course of Practical Physics*.

* This method will also apply with sufficient accuracy to the correction for the emergent stem if that correction is small.

Now turn *D* into position A, and set the pump working. When the difference of level of the mercury columns in *C* is 4 cms., turn the stop-cock into position C, turn off the water at tap T_1, and, after waiting two minutes, read the thermometer and manometer three times at half-minute intervals. After taking the observations, turn on the water again, place *D* vertical with the dot to the right, and diminish the pressure 4 cms. further. Take readings as before.

Repeat the observations at intervals of 4 cms. pressure till the temperature is about 75°; then with the pressure increasing again to that of the atmosphere. Take the mean of each set of three, correct it for the errors of the thermometer, and tabulate your results as follows:

Date:

Observed height of barometer... ... 763·2 mms.
Calculated boiling point ... ... 100°·1 C.
Observed reading of thermometer at atmospheric pressure ... ... 99·3°
Hence correction of thermometer at boiling point ... ... = + ·8°

Pressure		Temperature		
at manometer	in flask	Observed	Correction	Corrected
0	763·2	99·3	+·8	100·1
40·2	723·0	97·8	·8	98·6
78·1	685·1	96·3	·8	97·1
121·2	642·0	94·6	·8	95·4
162·1	601·1	92·8	·8	93·6
&c.	&c.			&c.

Now plot a curve with pressures as abscissae, and the corrected temperatures as ordinates, representing the points obtained with the pressure decreasing by a cross, those with the pressure increasing by a circle.

The two sets of observations should not differ by more than about ·1° from each other. If they do, sufficient time has not been allowed between each observation for the temperature of the steam to become constant.

If the agreement is good plot a second curve with log p as abscissae and temperatures as ordinates. Notice that this curve is nearly a straight line.

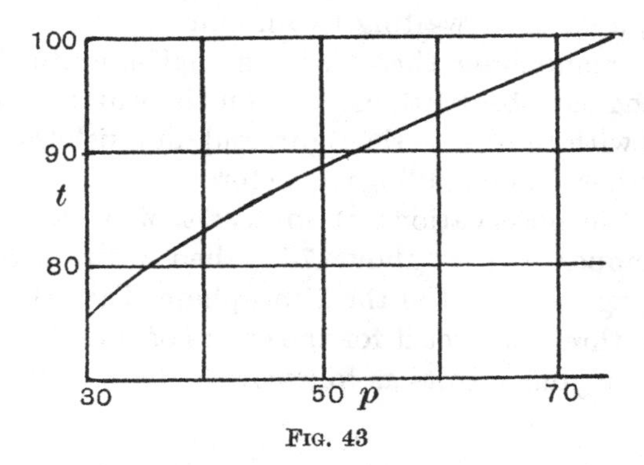

FIG. 43

From the observations determine the temperatures corresponding to pressures of 760, 720, 680 mms., &c., by means of the curves or as follows:

Let p_1 be the observed pressure next below the pressure p for which the temperature t of boiling is to be determined. Let p_2 be the observed pressure next above p. Then if t_1 and t_2 be the temperatures of boiling corresponding to p_1, p_2

$$t = t_1 + (t_2 - t_1)\frac{\log p - \log p_1}{\log p_2 - \log p_1} \text{ or } t = t_1 + (t_2 - t_1)\frac{p - p_1}{p_2 - p_1}$$

approximately.

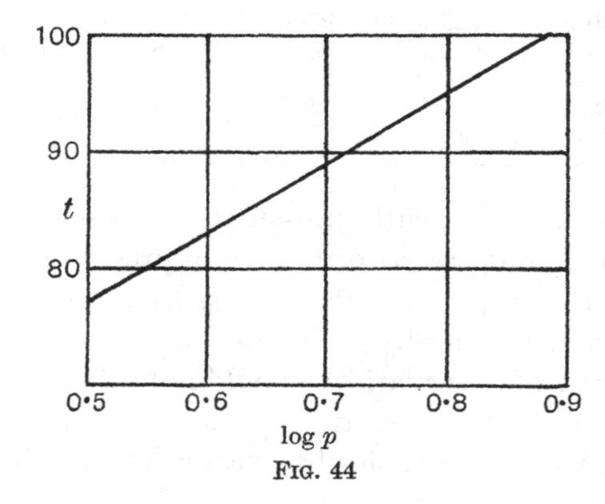

FIG. 44

The second term need only be calculated to one significant figure, so that the second expression is as a rule sufficiently accurate.

Tabulate the results along with Regnault's as follows:

Pressure	Temperature	
	Experiment	Regnault
760 mms.	100·0	100°·0 C.
720	98·6	98·5
680		96·9
640		95·2
600		93·5
560		91·7
520		89·7
480		87·6
440		85·4
400		83·0
360		80·4
320		77·5

SECTION XXI

LAWS OF COOLING

Apparatus required: *Calorimeter, with* (*a*) *simple enclosure,* (*b*) *enclosure with water jacket; glass tubing to be used as pipette; thermometer graduated in* $\frac{1}{10}$ *degrees and reading from about* 10° *to* 25° C.; *thermometer graduated in* $\frac{1}{10}$ *degrees and reading up to* 50° C.

If a body differs in temperature from its surroundings, it will lose or gain heat by radiation and by conduction to the bodies surrounding it.

The conduction is much affected by air currents caused by the air in contact with the calorimeter ascending and being replaced by colder air. Conduction is thus increased by the fresh supply of cold air, and is rendered irregular if by any cause the movement of the air is accelerated or retarded. Heat carried away by the bodily motion of the air is said to be carried by "convection."

Exercise I. To determine the rate of cooling of a calorimeter under various conditions.

Draw a pencil line round the inside of the calorimeter (Fig. 45 *a*) provided, about 1 cm. below the rim.

Fill it to the mark with water at a temperature about 20° above that of the room. Place the calorimeter on the table unprotected, and place a thermometer, reading to 50°, in the water. Take readings of the temperature of the water every half minute during a period of two minutes, *stirring the water well the whole time*, and, estimating to 0°·01 C., wait two minutes and then take five further observations at half-minute intervals. Take the temperature of the air by means of the smaller thermometer in the interval of waiting.

Raise the water to its original temperature, and repeat the experiments with the calorimeter protected by an outside vessel but not covered (Fig. 45); and thirdly, with the calorimeter covered.

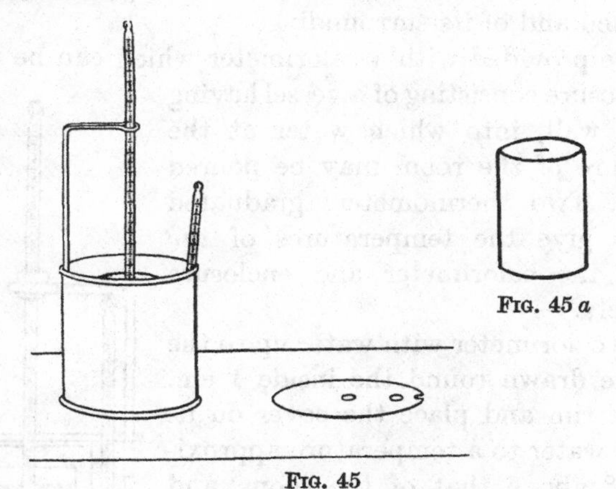

FIG. 45 a

FIG. 45

The experiments are entered and reduced as follows:

Date:

1. CALORIMETER UNPROTECTED.

Time hr. min.	Temperature	Time hr. min.	Temperature	Fall in 4 min.
3 10·0	38°·18 C.	3 14·0	37°·53 C.	·65
10·5	·10	14·5	·46	·64
11·0	·00	15·0	·37	·63
11·5	37·94	15·5	·29	·65
12·0	·86	16·0	·23	·63
Means	38·016		37·376	·640
Mean temp. 37°·696				

Average fall in four minutes ... ·640
,, ,, per minute ... ... ·160
Temperature of air ... ... 18°·5 C.

The rates of cooling when the calorimeter is protected by an outside vessel, and covered or uncovered, must be deduced in exactly the same way. Compare the results in the three cases, and draw your conclusions.

Exercise II. Having seen how the effects of convection currents can be diminished by properly protecting the calorimeter, we must next study how the amount of the cooling, in nearly quiescent air, varies with the temperatures of the body cooled and of its surroundings.

You are provided with a calorimeter which can be placed in an enclosure consisting of a vessel having a double wall, into which water at the temperature of the room may be poured (Fig. 46). Two thermometers graduated in tenths give the temperatures of the water in the calorimeter and enclosure respectively.

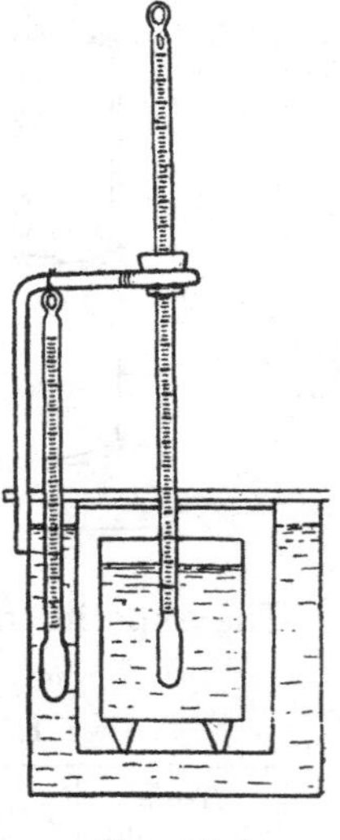

Fig. 46

Fill the calorimeter with water up to the pencil line drawn round the inside 1 cm. below the rim and place the cover on it. Raise the water to a temperature approximately 5° above that of the room, and observe the temperature every half minute for two minutes, then wait two minutes and observe again for two minutes, keeping the water well stirred. Note the temperature of the enclosure, which should be approximately the same as that of the room, at the beginning and end of the observations.

Tabulate as shewn previously.

Raise the water in the calorimeter to temperatures 10°, 20°, and 30° above that of the room, and observe the rate of cooling in each case, adjusting the temperature of the enclosure to be the same within a few tenths of a degree during each set of observations.

Calculate in each case the ratio of the rate of cooling to the excess of the mean temperature of the calorimeter over that of the enclosure. Call that ratio h'.

Collect your results as follows:

Mean temperature						
calorimeter	enclosure	Difference	Cooling per minute	h'	$(\text{difference})^{\frac{1}{4}}$	$\dfrac{h'}{(\text{difference})^{\frac{1}{4}}}$
48·54° C.	16·30° C.	32·24° C.	·494	·0153	2·38	·00643
39·31	16·25	23·06	·321	·0140	2·19	·00639
29·58	16·45	13·13	·158	·0120	1·90	·00628
21·66	16·40	5·22	·050	·0095	1·51	·00630

The quantity called h' is the rate of cooling for a difference of temperature of one degree between calorimeter and enclosure, calculated on the assumption that the cooling is proportional to the difference of temperature between the calorimeter and enclosure, an assumption which the result shews is only approximately true. When h' is divided by the fourth root of the difference of temperature the quotients h shew that the rate of cooling is proportional to the 5/4ths power of the temperature excess of the calorimeter over the enclosure. The mean value of h is ·00635.

If the mass of water in the calorimeter is M, then the heat lost will be Mh gramme degrees per minute, and if A is the total area of the calorimeter exposed, the heat lost per unit surface in one minute will be Mh/A, or if k is the heat lost per sq. cm. of surface per second, at 1° C. excess of temperature

$$k = \frac{Mh}{60A}.$$

Calculate k from the mean value of h found, and enter your results as follows. For small differences of temperature k should only depend on the nature of the radiating surface.

Inner radius of enclosure	6 cms.
Weight of water in calorimeter ...	146 grms.
Radius of calorimeter	3·2 cms.
Height of ,, ...	5·1 ,,
Area of curved surface of calorimeter ...	102·6 sq. cms.
,, top and bottom ,, ...	64·4 ,,
Total area of radiating surface	167·0 ,,

Mean value of h found above = ·00635.

Hence $$k = \frac{\cdot 00635 \times 146}{60 \times 167} = \cdot 0000924.$$

The inner radius of the enclosure is not required in the calculation, but should be measured and recorded, in order that the effect of the dimensions of the air space on the heat lost by conduction and convection may be traced by comparing the results obtained under different conditions.

Newton's Law of Cooling states that the rate of loss of heat of a hot body is proportional to the difference in temperature between the surface of the hot body and the surrounding space. Although the law is approximately true in the case of a body cooling in a current of gas it must be replaced by the $\frac{5}{4}$**ths power law** when the heated body produces the convection currents.

If h is the cooling in a given interval of time of a calorimeter containing a liquid whose specific heat does not vary with the temperature, when the difference in temperature between the calorimeter and the enclosure is one degree, the $\frac{5}{4}$ths power law states that the rate of cooling H when the temperature of the enclosure is T_0 and that of the calorimeter T is

$$H = h\,(T - T_0)^{\frac{5}{4}}.$$

If T_0 is known, the value of h may be determined from a single experiment, as in Exercise II. above.

It is often impossible to know the value of T_0 accurately, as different parts of the enclosure may have different temperatures. In this case the above equation will still hold, but T_0 will represent a mean temperature of the enclosure. It is then more convenient to write the equation

$$H^{\frac{4}{5}} = h^{\frac{4}{5}}\,(T - T_0).$$

Then if H_1 be the rate of cooling of the calorimeter when the temperature is T_1, and H_2 when the temperature is T_2, we have

$$H^{\frac{4}{5}} = \frac{H_1^{\frac{4}{5}}\,(T_2 - T) - H_2^{\frac{4}{5}}\,(T_1 - T)}{T_2 - T_1},$$

which gives the rate of cooling for any value of T.

Sufficiently accurate values may be obtained by a graphical method. Let OA_0 (Fig. 47) represent the temperature T_0 of the enclosure, and OA_1 the temperature T_1 of the calorimeter for which the cooling has been determined. Let A_1N_1 represent the $\frac{4}{5}$ths power of the rate of cooling measured on some suitable scale. Join A_0N_1. Then if OA be any temperature T, the corresponding $\frac{4}{5}$ths power of the rate of cooling will be measured by AN. For

$$\frac{AN}{A_1N_1} = \frac{AA_0}{A_1A_0} = \frac{T - T_0}{T_1 - T_0} = \left(\frac{\text{Rate of cooling at temperature } T}{\text{Rate of cooling at temperature } T_1}\right)^{\frac{4}{5}}.$$

Hence, if a curve is drawn such that the ordinates measure $\frac{4}{5}$ths powers of the rates of cooling, and the abscissae the corresponding temperature of the cooling body, the curve is a straight line within the limits of accuracy of the $\frac{5}{4}$ths power Law of Cooling.

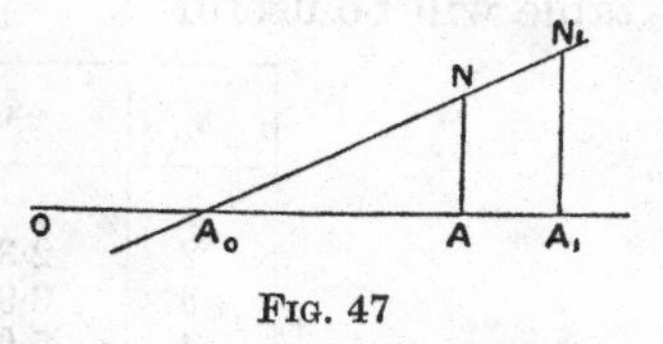

FIG. 47

If the temperature T_0 is not known, but the $\frac{4}{5}$ths powers of the rates of cooling H_1 and H_2 are found for the two temperatures T_1 and T_2, we may deduce the rate of cooling for a temperature T as follows:

Take OA_1 and OA_2 (Fig. 48) to represent T_1 and T_2, A_1N_1 and A_2N_2 the rates of cooling H_1 and H_2. Join N_1N_2. If OA be any temperature T, then AN will be the corresponding $\frac{4}{5}$ths power of the rate of cooling.

It may happen that at the beginning of an experiment, the water in the calorimeter at a temperature T_1 does not fall but rises in temperature. In that case the line A_1N_1 must be measured downwards as in Fig. 49, *i.e.*, it must be taken as negative cooling.

In drawing the curves, attention should be given to the scale. If the rates of cooling are measured to one per cent., for instance, which in all cases will be sufficient, the length

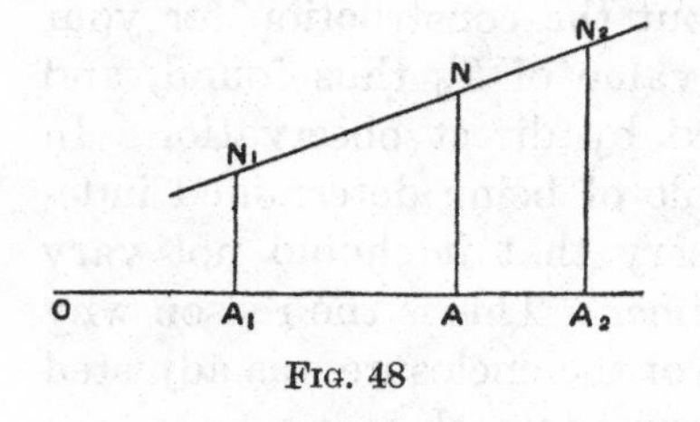

FIG. 48

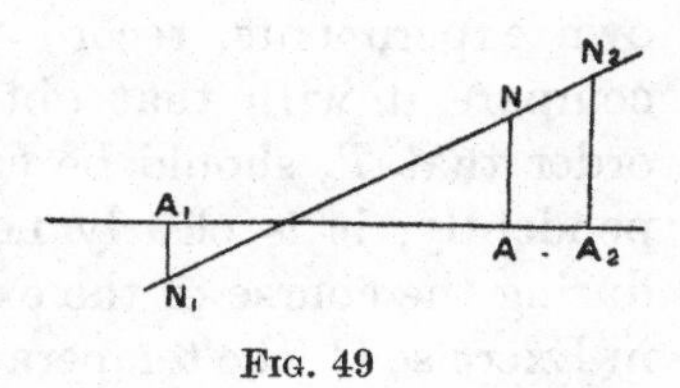

FIG. 49

of the line A_2N_2 should be such that the hundredth part of it may be estimated, *i.e.*, the unit in which it is measured should be chosen so that A_2N_2 is at least equal to ten times the length of a division on the squared paper used.

The scale of temperatures may conveniently be such that the angle between A_1A_2 and N_1N_2 is between 30° and 45°.

In the calculation of the $\frac{5}{4}$ths or $\frac{4}{5}$ths powers the following table will be useful.

u	$u^{\frac{5}{4}}$	u	$u^{\frac{5}{4}}$
1	1	11	20·03
2	2·38	12	22·33
3	3·95	13	24·69
4	5·66	14	27·08
5	7·48	15	29·52
6	9·39	20	42·30
7	11·38	25	55·90
8	13·46	30	70·20
9	15·59	40	100·60
10	17·79	50	132·95

A graph should be drawn so as to allow intermediate values to be read off directly.

From the results obtained in Exercise II., making use of the observed rates of cooling in the two cases in which the differences in temperature between the enclosure and calorimeter is least, determine, by the graphical method just described, the temperature of the enclosure, supposed not to be known. Thus, in the above example, one would take the abscissae proportional to 29·6 and 21·7 and the ordinates in some convenient scale proportional to 158 and 56. The line drawn through the two points so determined intersects the axis of temperatures at the point representing the temperature of the enclosure. Carry out the construction for your own experiments, record the value of T_0 thus found, and compare it with that obtained by direct observation. In order that T_0 should be capable of being determined independently, it is clearly necessary that it should not vary during the course of the experiment. This is the reason why in Exercise II. the temperature of the enclosure was adjusted so as to be approximately the same in each case.

Exercise III. Specific Heats of Liquids by Cooling.

Weigh the covered calorimeter, fill it with water to the mark and weigh again to determine the water added. Heat to about 32° and notice the time, t_0, it takes when enclosed by a water jacket at a definite temperature, say 17° C., to cool from say 30° C. to 28° C.

Fill the calorimeter to the mark with the liquid whose specific heat is required: weigh, heat to 32° and with the water jacket again at 17° notice the time, t_1, the liquid takes to cool from 30° to 28°.

The water equivalents of calorimeter and contents are proportional to the times taken. Hence if M_0 is the mass $C_0 = 1$, the specific heat of the first $M_1 C_1$ of the second m, c of the calorimeter itself,

$$\frac{M_1 C_1 + mc}{M_0 C_0 + mc} = \frac{t_1}{t_0}.$$

Determine the specific heat of carbon tetrachloride in this way.

SECTION XXII

COOLING CORRECTION IN CALORIMETRY

Apparatus required: *Calorimeter with water jacket, thermometer graduated in tenths of degrees, india-rubber and heater.*

This exercise is an application of the principles explained in the exercise on the "Laws of Cooling," to the determination of the specific heat of india-rubber. India-rubber is a bad conductor of heat and gives up its heat to the calorimeter slowly, in consequence the time occupied is considerable and the correction for cooling is of special importance.

The object of the exercise is not so much to obtain an accurate value of the specific heat as to shew the method of reducing calorimetric observations.

Place the india-rubber in a heater, and raise it to about 100° C. This may be done with sufficient accuracy by placing it in boiling water for not less than a quarter of an hour.

While the india-rubber takes up the proper temperature, weigh the calorimeter and fill it and the water jacket with water which has previously been standing in the room, so as to take up its temperature—tap water is generally much colder. Weigh the calorimeter again to determine the quantity of water in it.

Place the calorimeter with the thermometer in it in its water jacket, and leave it for two or three minutes before taking any observations. In the meantime prepare your notebook by drawing vertical lines to allow for five vertical columns. The first column is for the time at which an observation is taken, the second is for the temperature observed.

The experiment is divided into three stages, during which observations of temperature are taken every half minute.

1. The first or initial period, before the introduction of the india-rubber into the calorimeter. This should last six

minutes, and should only be begun when the change of temperature is steady and small. Observations are taken at half-minute intervals during two minutes, then stopped during two minutes and resumed again.

2. The second or principal period, in which the india-rubber is in the calorimeter. The thermometer will be found at first to rise, then to remain nearly steady, and then to fall. *When the fall has become uniform we arrive at:*
3. The third or final period, of six minutes, during which the fall is steady. Observations are taken as in the first period.

About 20 seconds after the last reading of the first period, the india-rubber is removed from the boiling water, the adhering water shaken off, and at the half minute the rubber is lowered into the calorimeter.

During the ten seconds that the india-rubber is exposed to the air the water on its surface evaporates, and is not carried into the calorimeter.

In an accurate measurement of specific heat this would not be a permissible process, as the parts of the india-rubber exposed to the air are cooled by evaporation of the water and by conduction to the air. The substance would have to be heated in an air chamber kept at 100° C. by being surrounded by steam, and be suddenly dropped into the calorimeter.

When the india-rubber has been placed in the calorimeter, the thermometer is observed at the next half minute and each succeeding half minute. The first few temperature observations will only be approximate, as the temperature will be rising quickly.

Observation of temperature should be made till it has ceased to rise and has fallen at a constant rate for six minutes.

In order to reduce the observations and calculate the specific heat, deduce from the observations taken during the first and third periods the rates of cooling or heating at the mean temperatures of the two periods, and determine graphically, as explained in the previous section, the rate of cooling corresponding to any temperature of the calorimeter*.

If the temperature rise is considerable it will be necessary to take as ordinates of the curves the $\frac{4}{5}$ths powers of the observed rates of cooling and having found from the curve the $\frac{4}{5}$ths power of the rate at any temperature to find from the table of powers p. 120 or from the graph of powers the actual rate of cooling at that temperature.

If the temperature rise is small the rate of cooling at any temperature may be found with sufficient accuracy by taking the rates of cooling themselves as the ordinates of the curve.

Enter in the third column of the note-book the mean temperature of the calorimeter during each interval of time throughout the experiment. Determine from the construction, and put down in the fourth column, the diminution of temperature owing to the loss of heat during each interval. In the fifth column tabulate the total loss in temperature up to each time given in the first column. An inspection of the table below will shew how this column is obtained. In the example given the loss was ·005 to the end of the first interval, and during the second interval the loss was ·009; the total loss up to the end of the second interval was therefore ·014, and if to this is added the loss during the third interval we obtain ·025 and so on.

The last column, which gives the temperature which the calorimeter would have shewn if there had been no loss of heat due to radiation and conduction to the air, is obtained by adding the numbers in the fifth column to those in the second.

It will be seen that the temperatures given in this column rise at first quickly, then slowly, and during the last period remain sensibly constant. The mean corrected temperature during that period would therefore have been 22°·54 if there had been no loss of heat.

* If the water equivalent w of the body placed in the calorimeter bears a large proportion to that of the calorimeter and contents W_1, and H_1 is the observed rate of loss of temperature in the first period, the quantity used in the graphical determination of the rate of loss at any temperature should be $H_1 \frac{W}{W+w}$ and not H_1.

FIRST PERIOD. Date:

Temperature slowly falling.

Time	Temp.	Time	Temp.	Cooling
$11^h\ 00{\cdot}0^m$	19°·69	$11^h\ 4{\cdot}0^m$	19°·67	·02
0·5	·69	4·5	·67	·02
1·0	·69	5·0	·66	·03
1·5	·68	5·5	·66	·02
2·0	·67	6·0	·66	·01
Means	19°·684		19°·664	·020

The mean of 19·684 and 19·664 gives the

Temperature at $11^h\ 3^m$ = 19°·674

Cooling in 4 mins. = ·020

" $\frac{1}{2}$ " = ·0025

" $3\frac{1}{2}$ " = 0°·018

Temperature at $11^h\ 6^m{\cdot}5$ when rubber introduced } = 19°·656.

SECOND PERIOD.*

	Temp.	Average temp. during Interval	Cooling in Interval	Total loss	Corrected temp.
11 h. 07·0 m.	20·85° C.	20·26	·005	·005	20·86
07·5	21·30	21·08	·009	·014	21·31
08·0	21·60	21·45	·011	·025	21·62
08·5	21·77	21·68	·012	·037	21·81
09·0	21·90	21·84	·013	·050	21·95
09·5	22·01	21·96	·014	·064	22·07
10·0	22·08	22·04	·014	·078	22·16
10·5	22·17	22·12	·014	·092	22·26
11·0	22·20	22·18	·015	·107	22·31
11·5	22·23	22·22	·015	·122	22·35
12·0	22·26	22·24	·015	·137	22·40
12·5	22·28	22·27	·015	·152	22·43
13·0	22·30	22·29	·015	·167	22·47
13·5	22·31	22·30	·015	·182	22·49
14·0	22·32	22·32	·016	·198	22·52
14·5	22·32	22·32	·016	·214	22·53
15·0	22·31	22·32	·016	·230	22·54

THIRD PERIOD.*

Temperature falling steadily.

	Temp.	Average temp.	Cooling in Interval	Total loss	Corrected temp.
11 h. 15·5 m.	22·30° C.	22·31	·015	·245	22·54
16·0	22·28	22·29	·015	·260	22·54
16·5	22·27	22·28	·015	·275	22·54
17·0	22·26	22·26	·015	·290	22·55
17·5	22·24	22·25	·015	·305	22·54
18·0	22·22	22·23	·015	·320	22·54
18·5	22·20	22·21	·015	·335	22·54
19·0	22·20	22·20	·015	·350	22·55
19·5	22·19	22·20	·015	·365	22·56
20·0	22·17	22·18	·015	·380	22·54
20·5	22·15	22·16	·015	·395	22·55
21·0	22·13	22·14	·014	·409	22·54
21·5	22·11	22·12	·014	·423	22·54

Mean during last period	22·544
Initial temperature ...	19·656
Rise of temperature ...	2·888

* See Note at the end of the Section.

Calculation of Rate of Cooling during third period.

22·30	–	22·19	=	·11
22·28	–	22·17	=	·11
22·27	–	22·15	=	·12
22·26	–	22·13	=	·13
22·24	–	22·11	=	·13
		Mean	=	·120

Hence cooling in 4 minutes = 0°·120

„ „ $\frac{1}{2}$ minute = 0°·015

This cooling corresponds to a mean temperature of 22°·2 which is obtained with sufficient accuracy from the temperature observed at the middle of the third period.

In an accurate determination of specific heats we are able to measure a rise in temperature of a few degrees to about one part in a thousand, although the observations have only been taken to the hundredth part of a degree. The increased accuracy is obtained by making use in the initial and final stage of a number of observations and taking the mean. In the present instance, the initial temperature of the india-rubber, owing to its treatment, must be doubtful to one degree, and the last figure need not be taken into account in the final calculations, although it should be worked out as in the example given for the sake of practice.

Knowing the masses of india-rubber and water, and the water equivalents of the thermometer and calorimeter*, the specific heat of india-rubber can now be calculated.

The final results are entered as follows:

Mass of india-rubber	= 12·64	grams.
„ calorimeter	= 28·14	„
„ „ and water	= 155·9	„
Mass of water	= 127·8	„
Calorimeter and thermometer equivalents ...	= 3·2	„
Total water equivalent	= 131·0	„
Initial temperature of india-rubber	= 100° C.	
Final „ „ „	= 22°·5	
Fall in „ „ „	= 77°·5	
Rise of temperature of water	= 2°·89	
Hence specific heat of india-rubber	= 0·387	

* The water equivalent of the calorimeter is the product of its mass into its specific heat.

The water equivalent of that part of the thermometer immersed in the calorimeter may be calculated from the volume immersed, since the heat capacities per unit volume of mercury and glass are nearly equal, *i.e.* ·45.

Draw curves (Fig. 50) shewing the observed changes of temperature, and the temperature curve after correcting for cooling.

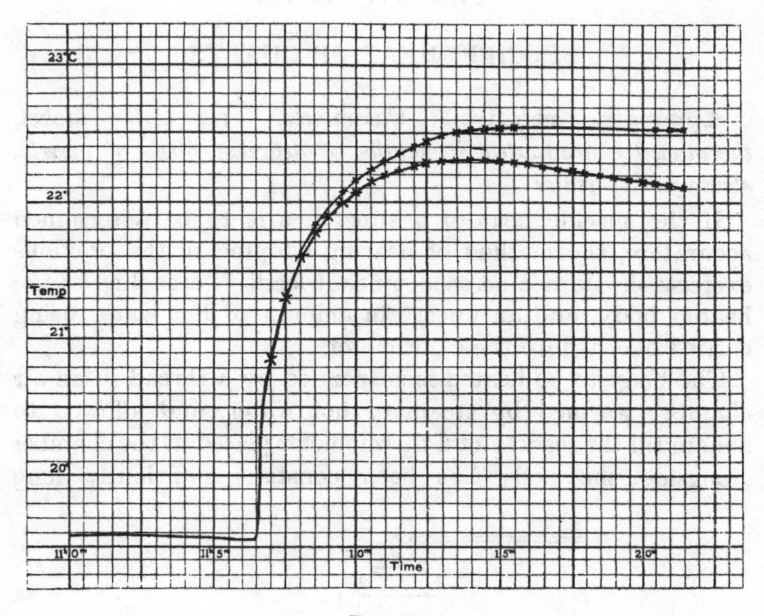

FIG. 50

The circumstances under which the loss of heat has been determined during the first and third periods are not strictly the same. At first the india-rubber was not immersed, and therefore the water equivalent of calorimeter and contents was different, and also the level of the water in the calorimeter changes. These introduce errors in the cooling correction, which are however very small when the change of temperature during the first period is small.

Note. When unnecessary decimals are discarded, it is an elementary rule to increase the last remaining figure by one when the first discarded figure is higher than five or five followed by other figures. But when only one figure is rejected, that figure being five, some doubt may arise as to whether to leave the last figure as it stands or to increase it by one. We should commit a systematic error if in the last column of the table on page 125 we were uniformly to take either the higher or the lower estimate. An excellent rule, adopted in the United States and deserving to come into general use, is always to leave the last figure an *even* number. Thus 25·35 and 30·65 should be shortened into 25·4 and 30·6.

SECTION XXIII

SPECIFIC HEAT OF QUARTZ

Apparatus required: *Calorimeter with water jacket, thermometer graduated in tenths of degrees, piece of quartz, steam-jacket heater.*

If the specific heat of a substance is to be determined accurately, the method of heating adopted in the previous exercise should not be used, since it leads to a cooling of the heated body, and an uncertain amount of hot water being carried over it into the calorimeter.

The body to be heated is suspended by a thread in an air chamber warmed by steam or hot water, and allowed to remain till the reading of a thermometer placed in the chamber alongside the body has been constant sufficiently long

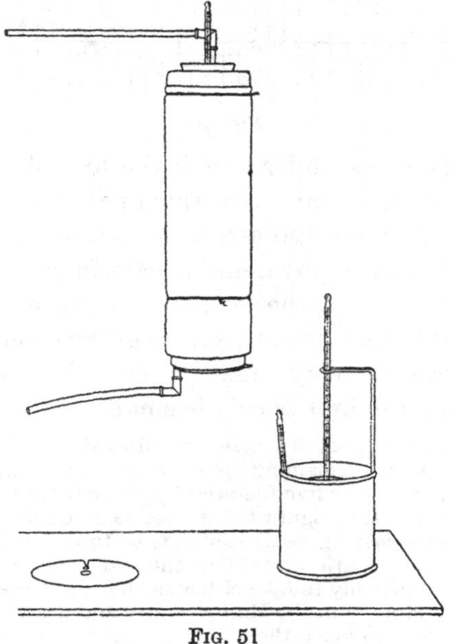

FIG. 51

to secure uniformity of temperature throughout the heated body.

Fasten about 50 cms. of thread to the piece of quartz provided, and suspend the quartz in the centre of a steam-jacket heater (Fig. 51).

Measure the diameter of the thermometer, and the length of its bulb and of that part of its stem which is immersed in the water of the calorimeter; hence calculate the volume to about ·5 c.c. Assuming it to be all mercury, calculate the water equivalent. As the heat capacities of equal volumes of glass and mercury are not greatly different, the equivalent of a thermometer may often be calculated in this way with sufficient accuracy. (See note p. 126.)

While the substance is heating fill the calorimeter about two-thirds full of water at the temperature of the room, the quantity of water used being found by weighing the calorimeter before and after filling. When the thermometer in the heater has been steady for 15 minutes, commence taking observations of temperature of the calorimeter for the "First period," as described in the previous section.

At the end of this period place the calorimeter and jacket under the heater and let down the quartz into the calorimeter as quickly as possible.

Replace the calorimeter in its original position, take observations during the "Second" and "Third" periods, and thence calculate the specific heat of quartz as in the previous section.

Calculate the specific heat also by the following approximate method: determine the rates of change of temperature per interval of time during the first and third periods, and thence deduce the rate of loss at that temperature which is the mean of the temperature at the commencement of the second period, and the *maximum* temperature attained.

Multiply this rate of loss by the number of intervals between the commencement of the second period and the time at which the maximum temperature is attained, and add the product to the maximum temperature. This will, if the time is short, *i.e.* if the substance gives up its heat rapidly to the water of the calorimeter, be approximately the temperature which the calorimeter would have attained if there had been no loss of heat.

Record the observations and calculations as in the previous section, and the additional calculation as follows:

Date:

Rate of cooling during first period = 0°·01 per half minute.
Temp. on dropping quartz in ... = 19°·72 C.
Maximum temp. attained ... = 22°·99
Interval between dropping quartz and max. temp. } = 3 half minutes.
Rate of cooling during third period = 0°·025 per half min.
Mean rate of cooling ... ... = 0°·0175 per half min.
∴ Loss of temp. ... = ·0175 × 3 = 0°·052
Corrected maximum temp. ... = 23°·04
Rise of temp. ... ... = 3°·32
Water heated + water equivalents = 181 grams.
∴ Heat absorbed by water ... = 601 gram-degrees.
Initial temp. of quartz ... ... = 99°·8
Final " " ... ... = 23°·0
Decrease of temp. ... ... = 76°·8
Weight of quartz... ... ... = 41·2 grams.

$$\therefore \text{Specific heat of quartz} = \frac{601}{41{\cdot}2 \times 76{\cdot}8} = {\cdot}191.$$

Repeat the experiment, using a little more water in the calorimeter. If the two results obtained are concordant take the mean as the final result.

It is instructive to compare the results obtained by this approximate method of treating the cooling with the more accurate one given in the last section. It will be found that even in the case of a substance taking up the temperature as quickly as quartz the difference is appreciable.

SECTION XXIV

LATENT HEAT OF WATER

Apparatus required: *Calorimeter, thermometer graduated in tenths of degrees, piece of ice.*

The latent heat of water, *i.e.* the heat absorbed when 1 gram of ice melts, may be determined by adding ice at 0° C. to sufficient water to melt the ice completely, and determining the decrease of temperature of the water.

Weigh the calorimeter provided, place in it about 170 grams of water at about 18° C., and weigh again.

Select a piece of ice weighing about 10 grams.

Place in the water a thermometer graduated to tenths of degrees, and observe the temperature of the water every half minute for 6 minutes ("First period") as explained in Section XXII.

Dry the surface of the ice thoroughly with blotting-paper, and at the end of the next half minute drop it into the water of the calorimeter. By means of the stirrer keep the ice under water and the water stirred. Take half-minute observations of temperature till it reaches a minimum and then commences to rise—"Second period." Continue observations of temperature for 6 minutes after the rise has become uniform, these observations constituting the "Third period."

From the observations taken in the first and third periods plot the curve from which the cooling at any temperature can be found, and apply the correction for cooling (or in this case heating) as described in Section XXII.

Weigh the calorimeter to determine the amount of ice which has been added, and from your observations determine the latent heat of water.

Work out from elementary principles the formula applicable in this case and record the observations and calculations as in Section XXII, and the additional calculations as follows:

Date:

Initial temperature of calorimeter ...	=	18°·48 C.
Final corrected ,, ,, ...	=	14°·24
Fall of temperature	=	4°·24
Weight of calorimeter and stirrer ...	=	59·11 grams.
,, cal. + water...	=	230·01 ,,
∴ ,, water	=	170·90 ,,
,, cal. + water + ice	=	238·25 ,,
∴ ,, ice added	=	8·24 ,,
Water equivalent of calorimeter ...	=	5·3 ,,
,, ,, thermometer ...	=	·5 ,,
Total water equivalent...	=	176·7 ,,

$$\therefore \text{Latent heat} = \frac{176{\cdot}7 \times 4{\cdot}24}{8{\cdot}24} - 14{\cdot}24 = 79{\cdot}1$$

Repeat the experiment, using a little more ice, and if the two results are nearly alike take the mean as the final result

SECTION XXV

LATENT HEAT OF STEAM

Apparatus required: *Calorimeter, condenser, thermometer graduated in tenths of degrees, flask, stand and burner.*

The latent heat of steam, *i.e.* the heat necessary to convert water at 100° C. into steam at the same temperature, may be determined by condensing a known weight of steam and observing the heat given up during the process to the liquid of the calorimeter, generally water, kept always below 100° C. The condensation takes place more regularly if the steam is allowed to condense in a separate vessel and not in the water of the calorimeter itself.

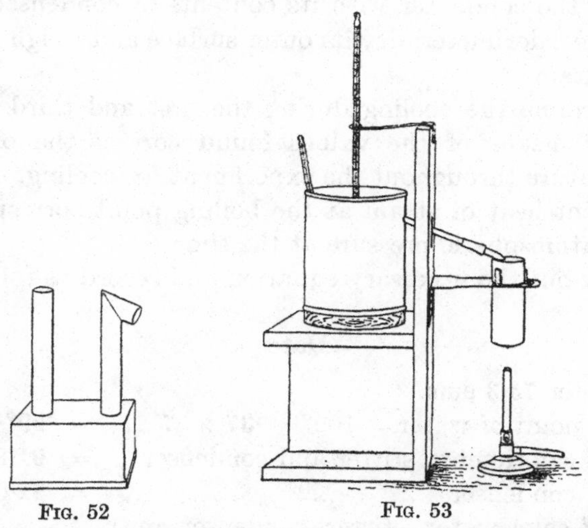

FIG. 52 FIG. 53

Weigh the calorimeter stirrer and condenser (Fig. 52) provided, first empty, then remove the condenser, weigh it to ·01 gram, replace, fill the calorimeter to within about 2 cms. of the top with water at the temperature of the room, and again weigh.

Arrange the vessel in which the steam is to be generated, the delivery tube and the calorimeter as shewn in Fig. 53.

When this has been done, disconnect the tube from the condenser and place a burner under the boiler, regulating the height of the flame so that the water boils gently, the steam being allowed to escape from the end of the tube.

Observe the temperature of the water in the calorimeter for 6 minutes—the "First period" of previous sections. Then replace the calorimeter, and insert the delivery tube into the head of the condenser tube so that the steam passes into the condenser and is condensed there. Keep the water well stirred, and observe the temperature every half minute, "Second period," till it has been raised about 10° C., then remove the delivery tube from the condenser.

Continue observations of temperature till the change of temperature has been uniform for 6 minutes, "Third period."

Weigh the calorimeter and contents, and for greater accuracy remove the condenser with its contents of condensed water from the calorimeter, dry its outer surface and weigh it again to ·01 gram.

Determine the cooling during the first and third periods and by means of the values found correct the observed temperature throughout the experiment for cooling, and find the latent heat of steam at the boiling point corresponding to the atmospheric pressure at the time.

Work out the necessary equation and record as follows:

Date:

Barometer 75·3 cms.

Boiling point of water = 100° − ·37 × ·7 ...	=	99°·74
Mass of calorimeter, stirrer and condenser	=	91·82 grms.
Mass of condenser	=	32·02 ,,
Mass of calorimeter, stirrer, condenser and water	=	223·24 ,,
∴ mass of water	=	131·42 ,,
Water equivalent of calorimeter, stirrer, condenser and thermometer	=	8·8 ,,
∴ total water equivalent	=	140·2 ,,

Initial temperature	14°·76 C.
Final temperature corrected according to method of Section XXI	32°·12
Rise in temperature	17°·36
Final weight of water, calorimeter, &c. ...	227·34 grms.
" " " condenser	36·13 "
Weight of steam condensed...	4·11 "

$$\text{Latent heat of steam} = \frac{17{\cdot}36 \times 140{\cdot}2}{4{\cdot}11} - 67{\cdot}6 = 532$$

Repeat the experiment, using a different quantity of water in the calorimeter, and if the two results are nearly alike take the mean as the final value.

Experiments on latent heat of vaporisation are liable to a number of errors owing to the difficulty of taking account of the gain and loss of heat at the point where the steam is led into the apparatus. Hence the results obtained when the experiments are made on a small scale are very uncertain. The above represents an average determination with the apparatus used.

SECTION XXVI

HEAT OF SOLUTION OF A SALT

Apparatus required: *Small calorimeter with suspending hook, larger calorimeters, thermometers, and salts.*

If p grams of a salt, the molecular weight of which is m, be dissolved in P grams of a solvent of molecular weight M, the solution formed has p/m gram molecules of the salt to P/M gram molecules of the solvent, or 1 gram molecule of the salt to every $n = Pm/pM$ gram molecule of the solvent.

If the specific heat of the solution formed be c, and if during the process the temperature of the solution decreases from t_0 to t, the quantity of heat absorbed by the solution of the salt is

$$\{(P + p) c + w\} \{t_0 - t\},$$

where w is the water equivalent of the calorimeter and thermometer. The quantity

$$q = \frac{\{(P + p) c + w\} \{t_0 - t\}}{p}$$

is the heat of solution of 1 gram of the salt, and the quantity

$$Q = m \frac{\{(P + p) c + w\} \{t_0 - t\}}{p}$$

is the heat of solution of 1 gram molecule of the salt, and is called the "molecular heat of solution."

The molecular heat of solution of a salt is nearly constant for weak solutions, but diminishes as a rule as the strength of the solution increases.

Determine the molecular heats of solution of Sodium Chloride and of Ammonium Chloride in water and their variations with concentrations by mixing

20 grams	NaCl ($m=58{\cdot}5$) in	98·5 grams	water ($M=18$),	$n=16$
15 ,,	,, ,,	,, 147·7	,, ,, ,,	$n=32$
5 ,,	,, ,,	,, 147·7	,, ,, ,,	$n=96$
15 ,,	NH_4Cl ($m=53{\cdot}5$) in	101	,, ,, ,,	$n=20$
10 ,,	,, ,,	,, 151	,, ,, ,,	$n=45$
5 ,,	,, ,,	,, 151	,, ,, ,,	$n=90$

The specific heats of the solutions may be taken as ·84, ·92, ·97, ·87, ·93 and ·96 respectively.

Proceed as follows:

Place the requisite quantity of water at about 18°·5 C. in a calorimeter surrounded by an air space and water jacket at the temperature of the room.

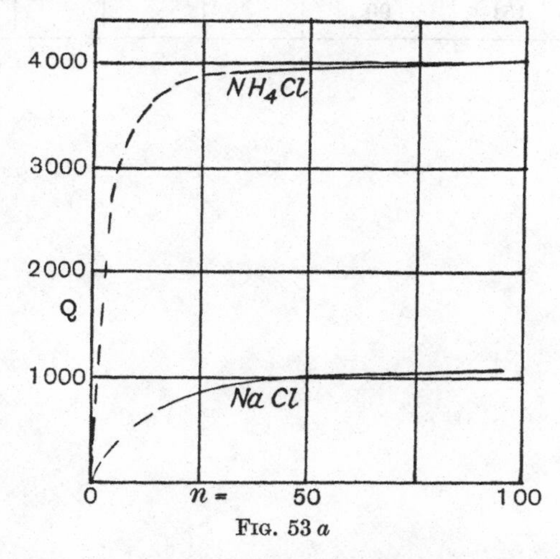

Fig. 53 *a*

Weigh the salt, put it into one of the small calorimeters, and suspend it by means of its hook in the water of the large calorimeter. Place a thermometer graduated to tenths of a degree in the water. After about 10 minutes take observations of temperature for 6 minutes. If the change of temperature is regular, unhook the small calorimeter and upset it in the water so that the salt and water come into contact with each other. Stir the mixture well and observe the temperature every half minute till the change has been regular for at least 6 minutes.

The solutions should not be left in the calorimeters.

From your observations recorded as on p. 125 make the

proper corrections for cooling, and determine the molecular heat of solution in each case.

Tabulate your results as follows,—and draw curves with n as abscissae, Q as ordinates, for each salt.

Salt	p	P	Molecules H_2O to 1 of salt $=n$	t_0	t observed	t corrected	q	Q
NaCl	20	98·5	16	18°·12	16°·14	16·05	11·4	660
,,	15	147·7	32	&c.			15·6	920
,,	5	147·7	96				18·3	1070
NH_4Cl	15	101	20	21°·32	13°·21	13·01	70·5	3800
,,	10	151	45	&c.			74·3	3970
,,	5	151	90				74·9	4010

SECTION XXVII

THE MECHANICAL EQUIVALENT OF HEAT OR SPECIFIC HEAT OF WATER IN WORK UNITS

Apparatus required: *Puluj's friction cones with rotating pulley, weights and hanger, jar of water, thermometer.*

When a gram degree of heat, *i.e.* the heat necessary to raise 1 gram of water at 15° C. to 16° C., is generated by the performance of mechanical work, the work done is called the mechanical equivalent of heat, or the specific heat of water in work units (ergs per degree).

To determine this quantity, the work may be done in a variety of ways; the one adopted in what follows depends on

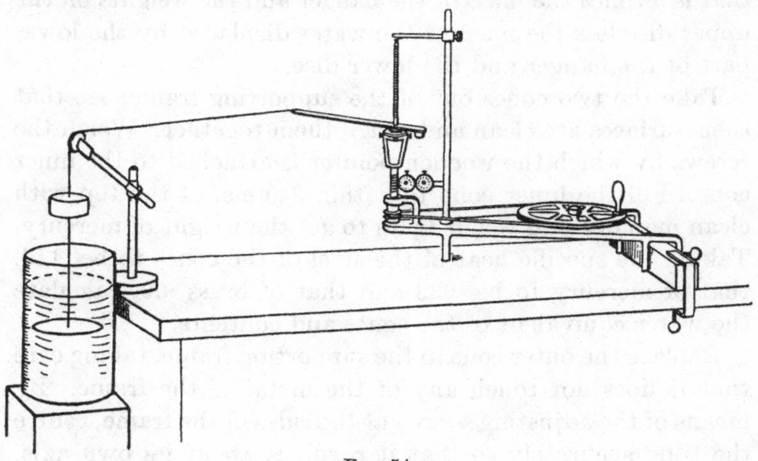

Fig. 54

one solid being made to slide along another against friction. In order that the sliding motion may be continuous the solids are circular, one, a small hollow cone of steel, fits into another similar cone slightly larger. The lower outer cone is held in, but thermally insulated from, a frame which can be set in rapid rotation about a vertical axis coincident with that of the cone. The smaller cone is filled with mercury and is placed in the rotating cone, but is prevented from rotating by a light wooden arm. To one end of this arm a thread is attached which passes over a pulley and carries a weight or a float placed in a jar of water.

The moment of the couple which the tension in the thread exerts on the inner cone is equal and opposite to that which the rotating outer cone exerts on the inner cone. The work done by the frictional couple in any interval of time is equal to this moment multiplied by the angle through which the outer cone has in the interval been rotated with respect to the inner cone. To determine this angle the apparatus is provided with two dial wheels which register the number of revolutions of the outer cone. The angle of rotation is 2π times the number of revolutions.

The tension in the thread is equal to the weight it supports, that is g times the mass of the hanger and the weights on the upper disc less the mass of the water displaced by the lower part of the hanger and the lower disc.

Take the two cones out of the supporting frame, see that their surfaces are clean and weigh them together. Weigh the screws by which the wooden pointer is attached to the inner cone. Fill the inner cone to within 3 mms. of the top with clean mercury and weigh again to get the weight of mercury. Taking the specific heat of the steel of the cones to be ·119, that of mercury to be ·033 and that of brass ·09, calculate the water equivalent of the cones and contents.

Replace the outer cone in the supporting frame, taking care that it does not touch any of the metal of the frame. By means of the adjusting screws at the sides of the frame, centre the cone accurately so that it revolves about its own axis. Attach the wooden pointer to the inner cone and place the cone in the outer as in the figure. Adjust the position of the dash pot and the length of the thread so that when the cone spindle is rotated at a convenient speed the thread and the wooden rod to which it is attached are perpendicular to each other. See that readings of the depth of the lower disc below the surface of the water can be taken on the graduated scale of the dash pot when it is raised by the thread. It should be possible to keep a weight of 50–100 grams suspended by the rotation of the cones by using a little petroleum as lubricant between the cones if the pull is too great, and a little vaseline if it is too small. Read the two dials attached to the rotating apparatus.

Measure the diameter and length of the bulb of the thermometer and calculate its water equivalent. Hang the thermometer on the movable arm attached to the stand and lower the bulb into the inner cone till it is below the level of the mercury.

Take observations of temperature every half minute for 3 minutes, then wait 3 minutes and take observations from 3 more minutes. At the end of this interval commence to rotate the hand wheel steadily, continuing to observe the temperature every half minute. At the middle of each half minute take a reading of the position of the disc in the water. Continue the rotation till the temperature has risen about 5° C. Then stop the rotation, read the temperature till the change has been regular for 3 minutes, wait 3 minutes, and read again for 3 minutes. Read the dials.

Remove the hanger from the water and determine its cross section and the diameter and thickness of the lower disc. Take also the mean of the readings of the lower disc and calculate the volume of the hanger in the water and therefore the mass of the water displaced. Subtract this from the mass of hanger and weights on upper disc. g times the difference is the mean tension in gravitation measure. Multiply this by 2π times the number of revolutions and by the length of the wooden arm and by g, and the product is the work done during the rotation in ergs.

Apply the correction for cooling to the temperature readings as in Section XXII and determine the corrected rise of temperature. The product of this by the water equivalent of cones and contents is the heat generated in gram-degrees, *i.e.* in terms of the unit of heat to which the specific heats used have been referred.

The quotient of the number of ergs of work done, by the number of gram-degrees of heat generated, is the number of ergs work required to generate heat sufficient to raise 1 gram of water 1° C.

Arrange your observations and results as follows:

Date:

Weight of steel cones ... ... = 84·4 grams.
,, ,, and mercury ... = 233·6 ,,
,, mercury ... ... = 149·2 ,,
Water equivalent of cones ... $84·4 \times ·116$ = 9·76
,, ,, mercury ... $149·2 \times ·033$ = 4·97
,, ,, thermometer ... ... = ·20
Total water equivalent of cones and contents ... = 14·93

Lower disc readings during rotation:

19·70, 19·60, 19·55... &c., mean ... ... = 19·52 cms.
Initial reading of counter... ... ... = 105
Final ,, ,, ... = 666
No. of revolutions ... ... = 561
Angle turned through $2\pi \times 561$ ... ... = 3532 radians.
Length of arm of lever ... ... = 25·7 cms.
Diameter of float stem = ·20, of disc ... = 8·0 ,,
Thickness of disc ... ... = ·05 ,,
Volume of hanger immersed ... ... = 3·12 c.c.
Mass of hanger and weights ... ... = 43·94 grams.
Tension ... ... = 40·82 g dynes.
∴ Work done = $40·82 \times 3532 \times 981 = 363 \times 10^7$ ergs.
Temperature at end of first period ... ... 15°·00 C.
Mean temperature at end of third period after correcting for cooling ... 20°·75
Rise of temperature ... ... 5°·75
Heat generated = $14·93 \times 5·75 = 85·8$ gram-degrees.

Mechanical equivalent = $\frac{363 \times 10^7}{85·8} = 42·3 \times 10^6$ ergs per degree.

Repeat the experiment, and if the two results are nearly alike take the mean as the final value.

BOOK IV

SOUND

SECTION XXVIII

FREQUENCY OF A TUNING FORK BY THE SYREN

Apparatus required: *Tuning fork, bow, singing flame, syren and blowing apparatus.*

To enable the comparison of a fork and a syren to be made more conveniently than it can be done directly, it is usual to tune a singing flame to the fork by adjusting the length and position of the resonating tube over the flame, and then to tune the syren to the flame.

Calculate the length of an open pipe which will act as a resonator to the fork, the vibration frequency of which is supposed to be known roughly. Take a glass tube of rather less length, and roll a piece of paper round one end so that the effective length of the tube may be altered by sliding the paper tube along. Place the tube above a small gas flame produced by gas issuing from a minute hole at the end of a glass tube drawn down fine and inserted in a Woolf's bottle connected to the gas supply.

Bring down the pipe on to the flame, and adjust till the flame "sings"; then vary the position of the paper tube till the note emitted by the pipe produces no beats with the note of the fork (Fig. 55).

Place the syren on the blowing apparatus, start the apparatus and increase the rate of blowing till the note emitted by the syren produces no beats with the pipe.

Maintain the rate of blowing, and at a given instant take a reading of the positions of the fingers on the dials indicating revolutions of the spindle of the syren.

At the end of a minute again take readings. Subtract the

readings to get the number of revolutions in the interval, count the number of holes in the disc of the syren, find the product and divide by 60, the result is the frequency of the note of the syren, and hence of the pipe, and fork.

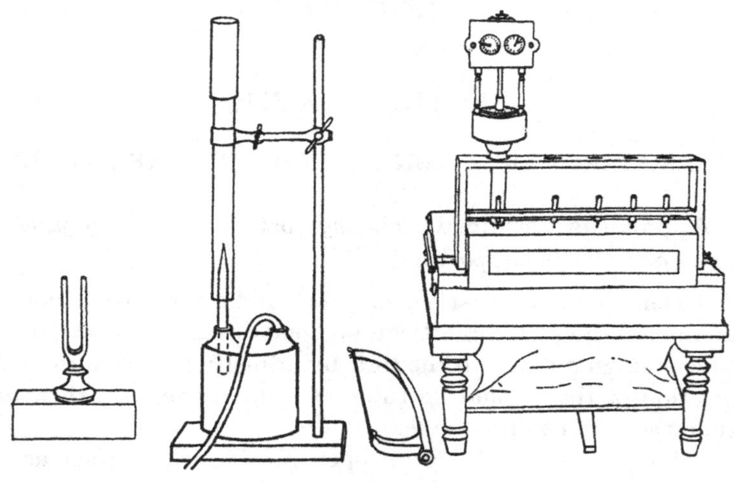

FIG. 55

Tabulate your results as follows:

Date:

Fork "A" marked 256.

Open pipe to act as resonator = 34000/512 = 66 cms.

Reading on syren dials at beginning of minute	300, 220, 410
,, ,, end ,,	910, 835, 1020
Differences	610, 615, 610
Mean	612

No. of holes in disc of syren = 25

∴ Frequency of fork ... = 255

SECTION XXIX

VELOCITY OF SOUND IN AIR AND OTHER BODIES BY KUNDT'S METHOD

Apparatus required: *Kundt's apparatus, rods and rubber.*

Kundt's apparatus consists of a glass tube of about 200 cms. length and 5 cms. diameter into one end of which a rod of wood, metal or glass projects (Fig. 56).

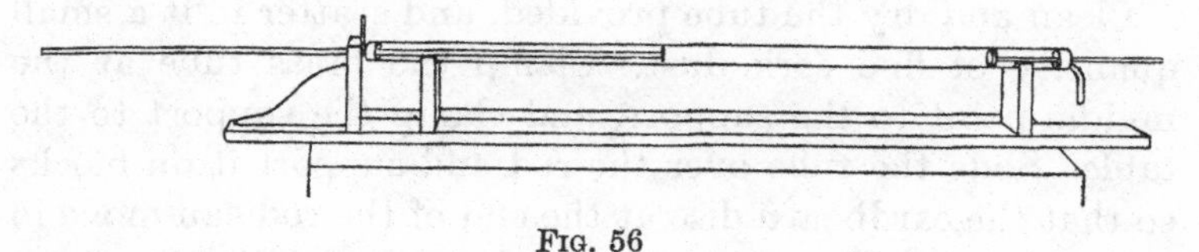

FIG. 56

The rod is securely clamped to the table at its middle point, and can be set into vibration parallel to its length by stroking it, if wood with a piece of cloth on to which a little resin has been rubbed, if metal with a leather rubber similarly treated, or if glass with a damp cloth. The end which projects into the tube is provided with a cardboard disc which has a diameter a little less than that of the inside of the tube.

The tube contains a little lycopodium powder or finely divided cork and the further end is stopped by a movable plug. It is important that both tube and powder should be quite dry.

When the rod is set into longitudinal vibrations, the disc on the end in the tube sets the air in the tube in vibration and the powder is carried along with the air. If the tube is long enough, there are certain parts of it where the air is not in motion along the tube, and at these parts the lycopodium or silica remains in heaps undisturbed.

These heaps therefore indicate the positions of the nodes of the vibrating column of air, and twice the distance between consecutive heaps is the wave length of the vibration in the gas with which the tube is filled—in this instance air.

Since the rod is clamped at its centre, this point will be a node in the rod, and the two ends will be the centres of vibrating segments, so that the wave length of the vibration in the rod is twice the length of the rod.

Since the frequencies of the rod and of the gas in the tube are identical, the ratio of the distance between the two heaps of lycopodium or silica to the length of the rod, is the ratio of the velocities of sound in the gas and in the material of the rod.

Determine by this method the velocity of sound in brass and in glass, given the velocity of sound in air, and compare the velocities of sound in air and in coal gas.

Clean and dry the tube provided, and scatter in it a small quantity of fine cork dust. Clamp the brass tube at the middle point to the support, and clamp the support to the table. Slide the tube over the rod and support it on blocks so that the cardboard disc at the end of the rod can move in the tube freely. Tap the tube sharply so that the cork collects in a line along the tube. Rotate the tube about its axis through a few degrees, so that the line of cork is not at the lowest point of the tube. Rub the rod lengthwise with a piece of leather on which resin has been rubbed, watching the line of cork during the process. If no motion is perceived, move the plug at the further end of the tube a centimetre and repeat. Continue till a position of the plug is found for which, when the rod is rubbed, some of the cork is blown along the tube and falls to the lowest point in a number of heaps. Move the plug by millimetres at a time till this action appears most energetic, and when the heaps are distinct and nearly touch each other count their number and observe on the scale under the tube the positions of three consecutive spaces between the heaps at each end of the tube. Take the mean of the first three and the mean of the last three, subtract and divide by the number of heaps between the mean readings, *i.e.* find the distance between consecutive heaps. Observe the temperature of the air in the neighbourhood of the tube, and measure the length of the brass rod. Substitute for the brass rod one of glass, and repeat the experiment, then one of wood and again repeat.

Record as follows:

Date:

Brass tube 129·5 cms. long.

Readings of centres of intervals between heaps of cork:

	30·00 cms.	91·50 cms.
	42·30	103·80
	54·70	116·20
Means	42·33	103·83

Difference = 61·50 cms. for 5 intervals.

∴ Distance of heaps apart = 12·30 cms.

Temp. of air ... ... = 18° C.

Velocity of sound in brass / velocity of sound in air

$$= 129{\cdot}5/12{\cdot}3 = 10{\cdot}53.$$

Velocity of sound in air at 18° C. = 341 metres per second.

∴ velocity of sound in brass = 3590 metres per second.

Similarly for the other rods.

By fitting corks to the ends of the tube and filling it with some other gas, the velocity of sound in that gas may be compared with the velocity in air. One of the corks should be attached to the clamp supporting the vibrating rod.

SECTION XXX

STUDY OF VIBRATIONS OF TUNING FORKS BY MEANS OF LISSAJOUS' FIGURES

Apparatus required: *Two large tuning forks, one the octave of the other, vibration microscope or lens and drop of mercury.*

When a tuning fork is set into vibration by one of the prongs being displaced from its normal position with respect to the other, both prongs are set into oscillation by the elastic forces which resist deformation of the fork. These forces depend mainly on the cross section of the prongs near their roots, while the masses which most influence the movement are those near the ends of the prongs. The influence on the frequency of a fork of a small mass added to a prong near one end may be readily studied by the help of "Lissajous' figures."

Arrange the two large tuning forks provided, so that the directions of vibration of the prongs are horizontal and vertical respectively. To the end of one prong of the fork vibrating vertically attach a small microscope or lens of 2 or 3 cms. focus, and to the other prong a small weight to counterbalance the lens. When this weight is in its proper position the vibrations of the fork continue longest for a given blow. Place the second fork in such a position that the outer end of one prong is seen through the lens distinctly (Fig. 57). Rub a little grease on the end surface of the prong, and rub a drop of mercury into the grease with the finger. The drop will break up into a number of smaller drops which will adhere to the surface. Move the fork till one of the drops is seen distinctly through the lens. Strike the prongs of the two forks with a rubber stopper in the hole of which a short wooden rod has been inserted to serve as a handle. On looking through the lens at the bead of mercury, a bright looped line is seen which changes its shape more or less quickly according to the relative frequencies of the two forks. The time which the curve takes to go through a cycle of

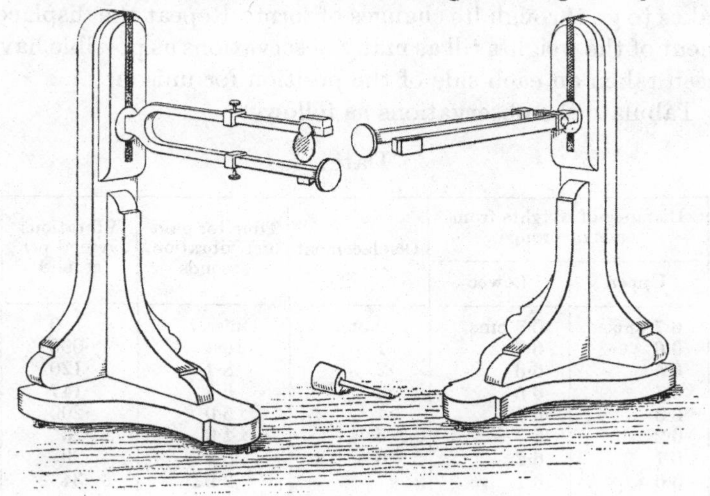

FIG. 57

changes should be approximately noted. The adjustable weights on the prongs of one of the forks should then each be moved slightly, say towards the ends of the prongs, and the observation of the time of a cycle repeated. If it is greater

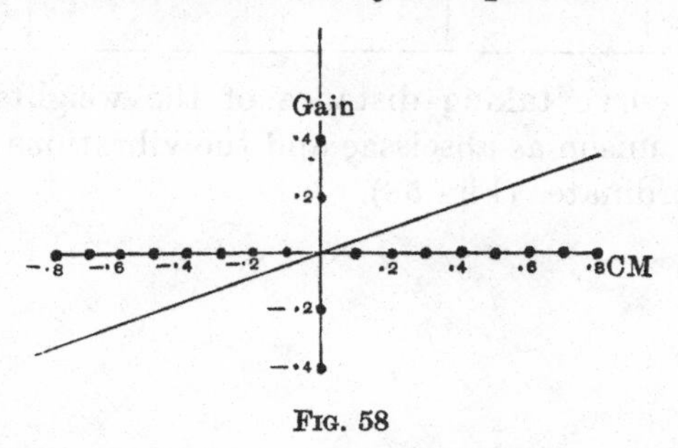

FIG. 58

than the previous time, the sliding weights should again be moved in the same direction, if less, in the opposite, till the time the curve takes to go through its changes of form is too long for the vibrations to be maintained throughout it. The two forks may then be considered in unison. Measure the distances of the weights from the ends of the prongs, move them both 1 mm. and observe the effect on the time the curve takes to go through its changes of form. Repeat the displacement of the weights till as many observations as possible have been taken on each side of the position for unison.

Tabulate the observations as follows:

Date:

Distance of weights from ends of prongs		Displacement	Time for gain of 1 vibration, seconds	Vibrations gained per second
Upper	Lower			
6·7 cms.	6·8 cms.	0 cm.	Unison	0
6·6	6·7	·1	10·4	·096
6·5	6·6	·2	8·4	·120
6·4	6 5	·3	6·8	·147
6·3	6·4	·4	5·0	·200
6·2	6·3	·5	4·0	·25
6·1	6·2	·6	3·48	·29
6·0	6·1	·7	2·92	·34
6·7	6·8	0	Unison	0
6·8	6·9	·1	10·0	·10
6·9	7·0	·2	8·6	
7·0	7·1	·3	7·2	
7·1	&c.			
7·2				
7·3				
&c.				

Draw a curve taking distance of the weights from the position of unison as abscissae and the vibrations gained per second as ordinates (Fig. 58).

BOOK V

LIGHT

SECTION XXXI

ANGLES BY THE OPTICAL METHOD

IF a body undergoes an angular displacement we may measure the displacement by reflecting a ray of light from a mirror attached to the body. If the plane of the mirror is parallel to the axis of rotation of the body, the reflected ray will turn through an angle which is double the angular displacement of the body.

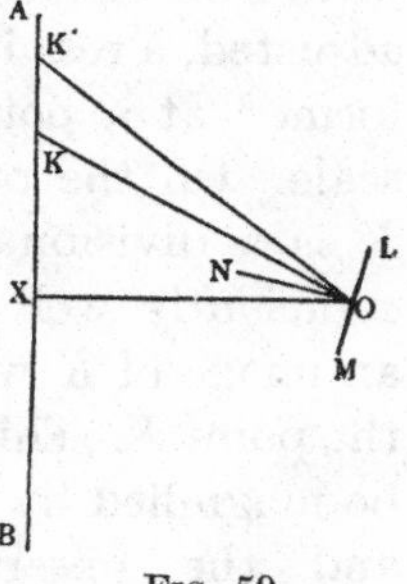

FIG. 59

In (Fig. 59) let XO be a ray of light incident on the mirror LM; let ON be the normal and OK the reflected ray, K being the point at which the reflected ray cuts a line drawn through X at right angles to the incident ray. If α is the angle between the incident ray and the normal, the angle $XOK = 2\alpha$ and hence

$$KX = OX \tan 2\alpha \text{ ....(1).}$$

If now the mirror turns through an angle θ so that $\alpha + \theta$ becomes the angle between XO and the normal, we shall have

$$K'X = OX \tan 2(\alpha + \theta) \text{ ....(2),}$$

where K' is the point at which the reflected ray cuts AB in the new position of the mirror.

Hence $$2\theta = \arctan\frac{K'X}{OX} - \arctan\frac{KX}{OX} \text{ ....(3).}$$

If the line AB is divided into equal divisions $K'X$ and KX can be read off directly and θ calculated from (3) after OX has been measured.

In order to make this method of measuring angles a practical one, we must of course deal with a beam of light and not with a single ray, and in that case the position of K will be well defined on the scale only when the beam comes to a focus at that point. In order to have the best definition the whole mirror should be filled with light, and hence the width of the beam at X must be twice the diameter of the mirror if it is plane.

Let in (Fig. 60) AB be the divided scale, L a lens with cross wires at F which can be illuminated from behind, then if the distance of the cross wires from the lens is properly adjusted, a real image will be formed at a point K of the scale. On the other hand if K is a division of the scale sufficiently well illuminated, an image of K will appear at the point F. This image may be magnified by an eyepiece, and the observer looking through what is now a telescope with objective at L and eyepiece at S, will see the divisions of the scale move through the field of view as the mirror turns round. The first method called "the objective method," in which the source of light is at F and the motion of the image at K is observed, is now very commonly employed. The apparatus is compact and the readings can be taken more quickly though not so accurately as with the other method. The objective method will be generally employed when, as in the case of a Wheatstone bridge, we do not wish to measure the deflections of a galvanometer mirror accurately, but only to make electrical adjustments until there is no deflection (null methods). It will also be employed when, as for instance in electrometer work, the unavoidable sources of error are so large that extreme accuracy in the reading of deflections would be waste of labour.

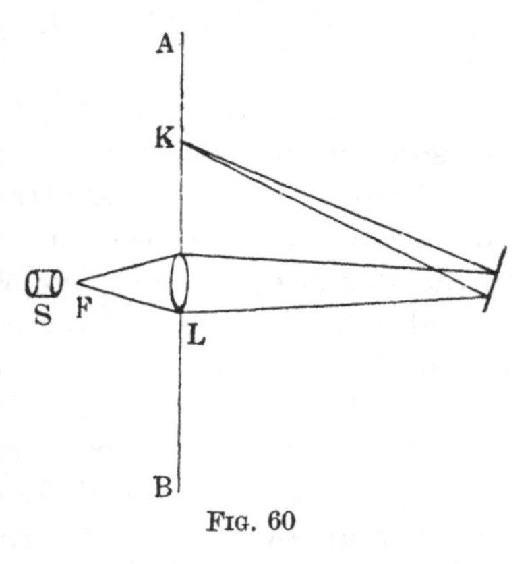

FIG. 60

The second or "subjective method" will generally be used when great accuracy is required; but it is necessary, in order to secure this end, that the galvanometer mirrors should not be too small and that the whole scale of the dimensions should be increased. This of course might also be done with the objective method, but other difficulties would then arise. The objective method requires that the scale should be properly shaded so that the spot of light may be easily seen; the subjective method requires that the scale should be properly illuminated so that its divisions should clearly appear in the telescope. The former condition is more easily secured when the apparatus is of small, the latter when it is of large dimensions. The large dimensions have the additional advantage that the observer works sitting at some distance from the instrument and hence irregular disturbances due to his moving about are often avoided.

In order to secure accuracy, it is important to adjust the position of the scale, and to apply a few corrections to the final result. We shall describe the adjustments for the case of instruments in which the subjective method is used, as they are then of special importance and may more easily be carried out. The student will have no difficulty in applying what is said to the objective method. We shall assume the mirror to turn round a vertical axis. The plan of the arrangement is that shewn in Fig. 60, but it is clear that the telescope must be placed so that its axis passes either above or below the scale. The scale, mirror and telescope will therefore all be at different levels and the telescope must either point downwards or upwards. This introduces no error in the result as long as OX in Fig. 59 is taken to be the horizontal projection of the optical axis of the telescope, as will be seen by considering that the image of each vertical scale division will also be vertical. If the telescope is therefore turned round a horizontal axis perpendicular to its length, the same division will always coincide with the cross wire, and though we may have to raise or lower the scale and mirror in a vertical plane in order that the image may appear in the field of view, no correction need be applied for want of horizontality of the optic axis. If the diameter of the mirror is equal to half that of the telescope, we can only obtain the maximum amount of light if the optic axis cuts the centre of the mirror, and in any case it is best to secure this for reasons of symmetry. The adjustment may be made by focussing the telescope on the mirror instead of on the image of the scale, moving it if necessary until the cross wires cut the mirror symmetrically; a little practice will however enable the observer to make the adjustment without altering the focus of the telescope, for its field of view is generally sufficiently large to see in addition to the scale a blurred image of the galvanometer parts immediately surrounding the mirror. If the clear image of the scale is seen in the centre of the field of view, the adjustment is sufficiently accurate.

In the equations (1), (2) and (3) it is assumed that the scale stands at right angles to the horizontal projection of the optic axis, and this adjustment must be made to the necessary degree of accuracy. This may easily be done, as the following investigation shews that the error introduced by a slight deviation from the correct position can readily be eliminated from the result.

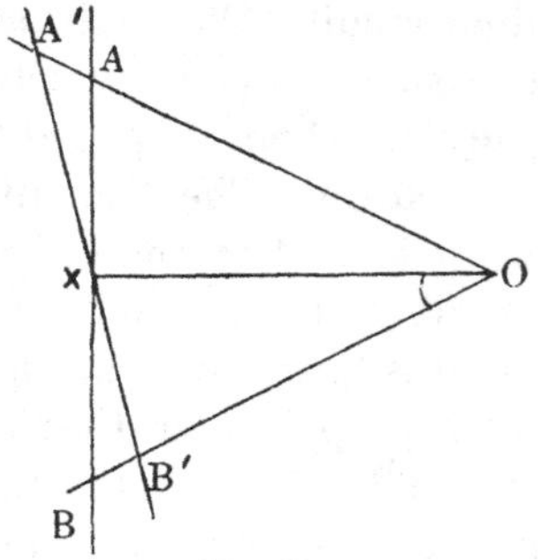

Let the projection of the optic axis cut the scale at X (Fig. 61) and the mirror at O, also let A and B be points on the scale equidistant from X, then if the scale is properly adjusted $OA = OB$.

A′ A X O B′ B

FIG. 61

The scale occupying the position $A'B'$, let γ be the angle BXB' and ϕ be the angle between OX and OB.

The triangle OXB' gives

$$\frac{B'X}{OX} = \frac{\sin\phi}{\sin\left(\gamma + \frac{\pi}{2} - \phi\right)} = \frac{\sin\phi}{\cos(\phi - \gamma)}$$

or

$$B'X = OX\,\frac{\tan\phi}{\cos\gamma + \sin\gamma\tan\phi} \quad \ldots\ldots\ldots\ldots(4).$$

The angle γ is supposed to be so small that its cube may be neglected compared to unity, thus

$$\cos\gamma = 1 - \tfrac{1}{2}\gamma^2, \quad \sin\gamma = \gamma,$$

and equation (4) becomes

$$B'X = \frac{OX\tan\phi}{1 + \gamma\tan\phi - \frac{1}{2}\gamma^2}$$

approximately.

If the scale had been placed in its proper position the line OB' would have cut it in B so that

$$BX = OX\tan\phi.$$

The observed reading should therefore be divided by

$$1 + \gamma\tan\phi - \tfrac{1}{2}\gamma^2$$

to give the correct reading.

If the mirror is deflected through the same angle but to the other side the observed reading should be divided by

$$1 - \gamma\tan\phi - \tfrac{1}{2}\gamma^2.$$

In actual experiments $\tan\phi$ is not likely to exceed $\frac{1}{4}$, and if the readings on either side are to be correct to 1 part in a thousand we must have $\frac{1}{4}\gamma + \frac{1}{2}\gamma^2 < \cdot 001$ or $\gamma < \cdot 004$, that is $< \frac{1}{4}^\circ$ and γ^2 may be neglected.

Since the triangle OXB' also gives

$$OB' = OX \cdot \frac{\cos\gamma}{\cos(\phi - \gamma)}$$

we may write

$$OB' = \frac{OX}{\cos\phi + \gamma\sin\phi}.$$

Similarly $$OA' = \frac{OX}{\cos\phi - \gamma\sin\phi}.$$

Hence $$OA' - OB' = OX\frac{2\gamma\sin\phi}{\cos^2\phi - \gamma^2\sin^2\phi},$$

and when $\tan\phi = \frac{1}{4}$ the difference $= OX \cdot \cdot 53\gamma = OX \cdot \cdot 0021$. If OX were 100 cms. the difference would be more than 2 mms.

The two ends of a scale 50 cms. long whose centre is 100 cms. from the mirror must be at equal distances from the mirror to within 2 mms. if the error of reading is to be less than 1 part in a thousand.

If the mean of the two readings one on each side of X is taken the error will only be dependent on the square of the angle γ through which the scale is turned, and the adjustment of the scale may be made with less accuracy to ensure the same accuracy of reading or with the same accuracy of adjustment we may secure a greater accuracy of reading.

To obtain the angle of deflection we have by formula (1) (Fig. 59)

$$2\alpha = \arctan\frac{KX}{OX},$$

which enables α to be found from tables of tangents.

If α is small it may also be calculated from the formula

$$\alpha = \frac{x}{2d}\left\{1 - \frac{4}{3}\left(\frac{x}{2d}\right)^2\right\},$$

where $x = KX$ and $d = OX$.

It is often necessary to find $\tan\alpha$, $\sin\alpha$, or $\sin\frac{\alpha}{2}$, instead of α, and the following approximate formulae may then be employed:

$$\tan\alpha = \frac{x}{2d}\left\{1 - \left(\frac{x}{2d}\right)^2\right\},$$

$$\sin\alpha = \frac{x}{2d}\left\{1 - \frac{3}{2}\left(\frac{x}{2d}\right)^2\right\},$$

$$\sin\frac{\alpha}{2} = \frac{x}{4d}\left\{1 - \frac{11}{2}\left(\frac{x}{4d}\right)^2\right\}.$$

In the above investigation it has been assumed that the normal to the mirror is at right angles to the axis of rotation, and this is sometimes difficult to secure with great accuracy. Calculation shews however that when the deviation of the normal does not exceed one degree the results are sufficiently accurate for nearly all purposes, provided that the distance d is measured from the centre of the mirror along the line which stands at right angles to the plane of the scale, to which the axis of rotation is supposed to be parallel. If as is commonly the case the axis of rotation is vertical, d should be the horizontal distance between the centre of the mirror and the plane of the scale.

Should it be necessary in exceptional cases to take account of the inclination of the mirror, this can be done by adding $x\alpha(\gamma - \alpha)$ to the observed deflection x, where α and γ are the angles which a plane drawn at right angles to the axis of rotation forms with the normal to the mirror and the optical axis of the observing telescope respectively. The angles α and γ must of course be measured towards the same side of the plane*.

* For other corrections see Holman, *Technology Quarterly*, Sept. 1898, or Czermak, *Reduktionstabellen zur Spiegelablesung*.

SECTION XXXII

THE SEXTANT

Apparatus required: *Sextant, fixed marks.*

The sextant consists of a graduated arc of a circle *AB* (Fig. 62), of about 60°, and a movable arm *I*, which turns

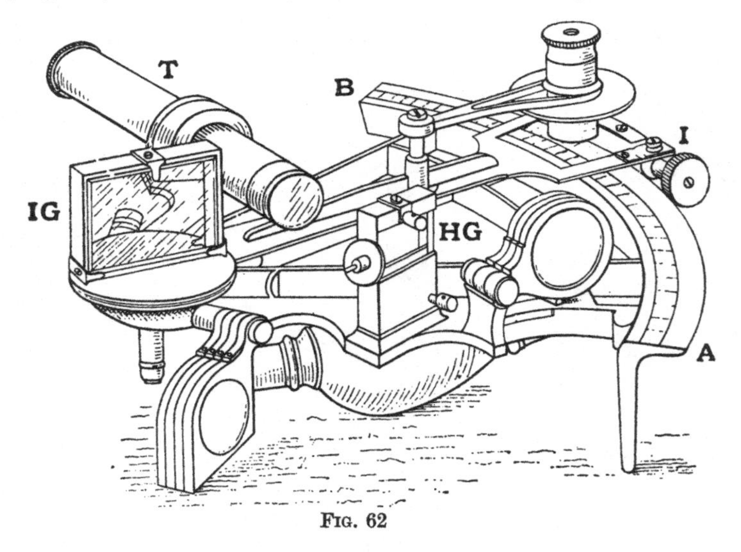

Fig. 62

about the centre of the arc, and is fitted with a clamp, a tangent screw, and a vernier by means of which its position on the graduated arc can be accurately determined.

IG is a plane mirror attached to the arm *I*, and is called the Index Glass. *HG* is a second mirror fixed to the frame, and known as the Horizon Glass. Its upper half is left unsilvered.

T is a small telescope directed towards the mirror *HG* and placed parallel to the plane of the arc. By means of a screw at the back of the instrument the telescope can be moved at right angles to this plane so as to vary the proportion of light received from the silvered and unsilvered portions of the horizon glass. The horizon glass is so placed that a ray of light passing from the centre of the index glass to the centre of the horizon glass is after reflection directed along the axis of the telescope (Fig. 63). Let such a ray come originally from an object Q, and let another ray coming from a second object P pass through the unsilvered part of the horizon glass and proceed in the same direction. The two objects when viewed through the telescope are then seen to coincide, one

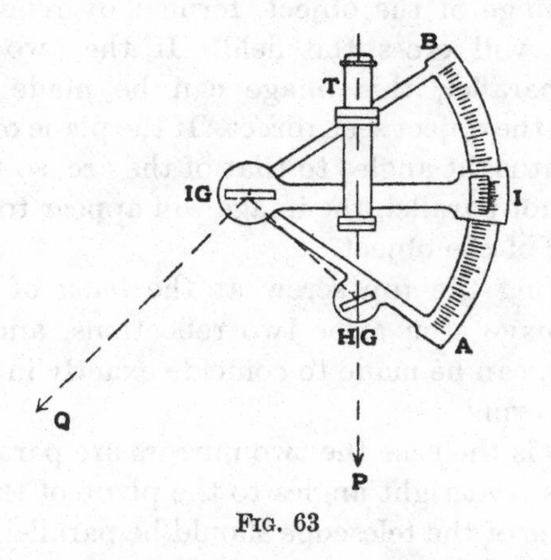

Fig. 63

being viewed direct and the other after reflection from the two mirrors, and the angle between the two mirrors, when this is the case, is half the angle between the rays from the objects.

The graduations on the arc are numbered to read double their real value; hence the reading on the arc gives directly the angle between two lines one of which is drawn from the centre of the index glass to Q and the other from the centre of the horizon glass to P. If the distance of the objects is sufficiently great this will be sensibly the same as the angular distance between P and Q at the observer's eye.

Adjustments.

1. The plane of the index glass IG should be at right angles to that of the graduated arc.

Verify that this adjustment is correct by setting the index at about 100°, placing the eye near the index glass, and looking obliquely at the glass so as to see at the same time part of the arc direct, and also its reflection in the glass. If the two appear to be in the same plane, the adjustment is correct, and the adjusting screws of the index glass need not be altered.

2. The plane of the horizon glass *HG* should be at right angles to that of the arc.

Hold the instrument so as to view directly with the telescope some small distant object. On turning the index arm round, an image of the object, formed by reflection at the two glasses, will cross the field. If the two glasses are accurately parallel, this image can be made to coincide exactly with the object seen direct. If the plane of the horizon glass is not at right angles to that of the arc, so that the two mirrors are not parallel, the image will appear to pass on one side or other of the object.

By adjusting the top screw at the back of the horizon glass, the image seen after two reflections, and the object seen directly, can be made to coincide exactly in one position of the index arm.

When this is the case the two mirrors are parallel, and the horizon glass is at right angles to the plane of the arc.

3. The axis of the telescope should be parallel to the plane of the graduated arc.

To test whether this condition is fulfilled with sufficient accuracy, place two small sights of exactly equal height on the divided circle. These sights are conveniently made by bending strips of sheet brass into two parts at right angles to each other, so that when resting on the sextant their upper edges are horizontal and at the same level as the axis of the telescope, the sextant being placed on a horizontal table. If the sights are placed as far apart from each other as possible and in suitable positions, the same distant object may either be viewed through the telescope or by the naked eye looking over the sights. A point on the distant object which may be brought into contact with the upper edges of the sights when looked at by the naked eye should then appear in the centre of the field of view. For accurate work with the sextant the adjustment of the telescope should be correct to within 10 minutes of arc or ·003 in angular measure. Hence if the sights are placed at a distance of 15 cms. from each other, they should be carefully constructed so that the two upper edges are horizontal and within half a millimetre of the same height above the divided circle. If the adjustment is found wrong it may be corrected in well-made sextants by two small screws in the frame which carries the telescope.

Determination of the Index Correction.

When the two mirrors are parallel, it may be found that the Vernier index does not read exactly zero. The reading is termed the index error, and the index correction is the index error with the sign reversed.

To determine the index correction, direct the telescope to a distant object and turn the index arm till the two images of the object appear in the field of view. Then clamp the index arm, and bring the images into coincidence by means of the tangent screw. Read the Vernier, and notice whether the reading is positive or negative.

The reading of the Vernier with the sign reversed is the index correction.

Exercise.

Standing on the given spot, determine the angle subtended by each pair of three given points.

Hold the sextant in the right hand, with the arc downwards. Look through the telescope at the lower one of two of the objects, holding the sextant so that the plane of the arc passes through both objects. Move the index arm along the arc till both objects appear at the same time in the field of view, then clamp the arm by means of the screw behind the scale, and bring the two images into coincidence by means of the tangent screw. Read the position of the index, repeat the observation, and record as follows. Proceed in the same way with each other pair of objects.

Date:

The reading when the two images of the same distant object coincided was found to be as a mean of 3 observations $-2' 12''$.

Readings when the images of the given points coincided:

Points:	Left and Middle	Middle and Right	Left and Right
	11° 30′ 20″ 30′ 30″ 30′ 00″	20° 3′ 10″ 3′ 0″ 3′ 20″	31° 35′ 20″ 35′ 40″ 35′ 40″
Means Index Correction	11° 30′ 17″ + 2′ 12″	20° 3′ 10″ 2′ 12″	31° 35′ 33″ 2′ 12″
Angle subtended	10° 32′ 29″	20° 5′ 22″	31° 37′ 45″

The observed points were nearly in a straight line.

It will often be found difficult to find suitable objects at a great distance. If the distance is less than 1000 yards, and the measurements of different observers are to be compared with each other, care must be taken that the observer does not change his position during the observations and that the position is well marked. Students will get a good idea of the delicacy of the observations by observing two objects at a distance of 1000 yards and at an angle of not less than 10° from each other; receding from or approaching the objects by a yard ought to make an appreciable difference in the coincidence of the images.

It would be difficult to name a more useful or instructive instrument than the sextant, and, if time allows, students should obtain a little practice in the determination of latitude and local time by means of it. The study of errors introduced by imperfections in the adjustment of the different parts will form an excellent foundation for the study of other optical instruments.

Consult: Chauvenet, *Spherical and Practical Astronomy*, vol. II.

SECTION XXXIII

CURVATURES AND POWERS OF LENSES

Apparatus required: *Two large lenses, one concave, one convex, one small convex lens, spherometer, scale, plane mirror, and simple optic bench.*

Curvature.

The radius of curvature of a spherical surface may be found by means of a spherometer, or by using the surface as a mirror, and determining the positions of conjugate points. Both methods will be used in what follows.

Place the spherometer provided on a plane surface, *e.g.*, a sheet of glass, and turn the milled head till the point of the screw just touches the surface. The exact point may be known either by the slight increase of the resistance to rotation of the screw, or by the rocking of the spherometer on the surface as soon as the point of contact is passed.

Take three readings of the zero of the spherometer in this way, and then place it on the surface the curvature of which is required, in this case a lens mounted in a ring to prevent the curved surface being scratched when the lens is placed on the table.

Again turn the screw till the point just touches the surface, making the adjustment three times and taking a reading each time.

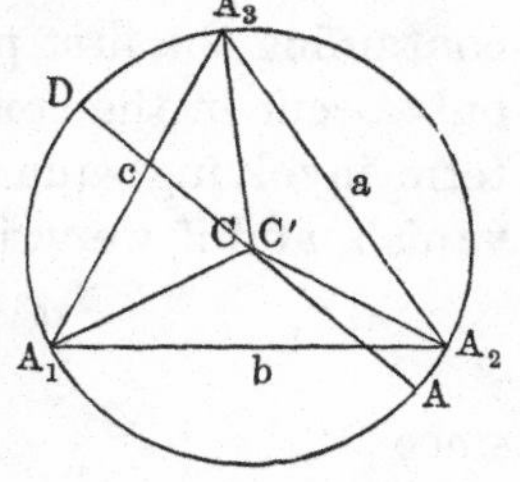

FIG. 64

The pitch of the screw is marked on the spherometer.

In order to determine the radius of curvature of the surface from these observations, we require to know the distances between the feet and point of the spherometer. Place the spherometer on a sheet of paper, and press it down gently, so that the positions of the feet and point are indicated by slight depressions A_1, A_2, A_3, C', in the paper (Fig. 64). Measure by means of a scale with its graduations in contact with the paper or by means of compasses, the distances A_2A_3, A_3A_1, A_1A_2, $C'A_1$, $C'A_2$, $C'A_3$, call them a, b, c, a', b', c'. The radius of the circle circumscribing the triangle $A_1A_2A_3$ will be given by

$$r = \frac{abc}{4\sqrt{s(s-a)(s-b)(s-c)}},$$

where $$s = \frac{a+b+c}{2}.$$

If $a = b = c$ this reduces to

$$r = \frac{a}{\sqrt{3}} = \frac{2s}{3\sqrt{3}}.$$

In practice a, b, c will not be exactly equal but nearly so, and in that case we may use the second form of the last equation.

In order to see within what limits of accuracy this is allowable put $a = \frac{2}{3}s + \alpha$, $b = \frac{2}{3}s + \beta$, $c = \frac{2}{3}s + \gamma$, where α, β and γ are small, then by addition

$$a + b + c = 2s + \alpha + \beta + \gamma,$$

or since $$a + b + c = 2s,$$

it follows that $$\alpha + \beta + \gamma = 0.$$

On substituting we now get

$$r = \frac{2}{3\sqrt{3}} \cdot \frac{(s + \frac{3}{2}\alpha)(s + \frac{3}{2}\beta)(s + \frac{3}{2}\gamma)}{\sqrt{s(s-3\alpha)(s-3\beta)(s-3\gamma)}}.$$

If we expand the products we find that there is no term containing the first power of α, β, γ because these quantities only occur in the combination $\alpha + \beta + \gamma$, which is zero. The term involving squares and products of α, β and γ does not vanish, and if we write

$$t^2 = \frac{1}{2}(\alpha^2 + \beta^2 + \gamma^2)$$
$$= -(\beta\gamma + \gamma\alpha + \alpha\beta),$$

since $$(\alpha + \beta + \gamma)^2 = 0,$$

it is easily found that to this degree of approximation

$$r = \frac{2s}{3\sqrt{3}}\left(1 + \frac{9}{4} \cdot \frac{t^2}{s^2}\right).$$

As an accuracy of one part in 500 will probably not be reached in this exercise unless the mean of a large number of determinations is taken, we may write

$$r = \frac{2}{3} \cdot \frac{s}{\sqrt{3}},$$

whenever t/s is less than ·031. If the second term cannot be neglected it is easily computed with sufficient accuracy.

If the spherometer were accurately made, the centre of the screw would move along a line through the centre C of the circumscribing circle $A_1A_2A_3$, perpendicular to the plane $A_1A_2A_3$. If it does not pass through C but through a point C' distant a', b', c' from the points $A_1A_2A_3$, then the distance CC' $(= d)$ is given approximately by the equation

$$d^2 = \tfrac{4}{9}\{2\left(\overline{r-a'}^2 + \overline{r-b'}^2 + \overline{r-c'}^2\right) + \overline{r-a'}\,\overline{r-b'} + \overline{r-a'}\,\overline{r-c'} + \overline{r-b'}\,\overline{r-c'}\}.$$

Now consider a section of the surface by a plane through O and the centre of curvature of the surface $C'C$, *i.e.* the circle $ANDN'$ (Fig. 65). The plane through the points $A_1A_2A_3$ is represented by the line ACD.

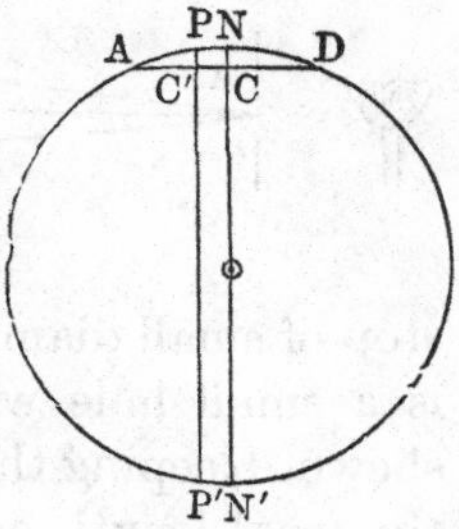

Fig. 65

Assuming that when the instrument is placed on a horizontal surface the axis of the screw is vertical, the distance h through which the point of the screw has to be moved when the instrument is placed on the spherical surface, will be $C'P$.

Now $C'P.C'P' = C'A.C'D$, and if $ON = R$,

$$PP' = 2\sqrt{R^2 - d^2} \text{ and } C'P' = 2\sqrt{R^2 - d^2} - h.$$

Hence $\qquad h\,(2\sqrt{R^2 - d^2} - h) = (r - d)\,(r + d),$

$$R^2 - d^2 = \frac{(r^2 + h^2 - d^2)^2}{4h^2},$$

$$R = \frac{1}{2h}\sqrt{(r^2 + h^2 - d^2)^2 + 4d^2h^2},$$

or if d is small, $\qquad R = (r^2 + h^2)/2h.$

Calculate R from one of these equations, and determine similarly the radius of curvature of the other surface of the lens.

Verify by the following optical method the values found:

I. For a Concave surface.

Place a screen provided with the hole and cross wires at one end of the optic bench with the white side towards the centre of the bench. Place a bright flame behind the screen. Place the lens on the bench with its axis parallel to the axis of the bench, and the surface the curvature of which is to be measured towards the screen. Find the position of the surface when the image of the cross wires on the screen produced by reflection at the surface is most distinct. The cross wires are then at the centre of curvature of the surface.

II. If the surface is convex the radius of curvature may be determined optically by the following method:

C (Fig. 66) is the lens the radii of curvature of which we have to determine. B is an auxiliary lens provided with a

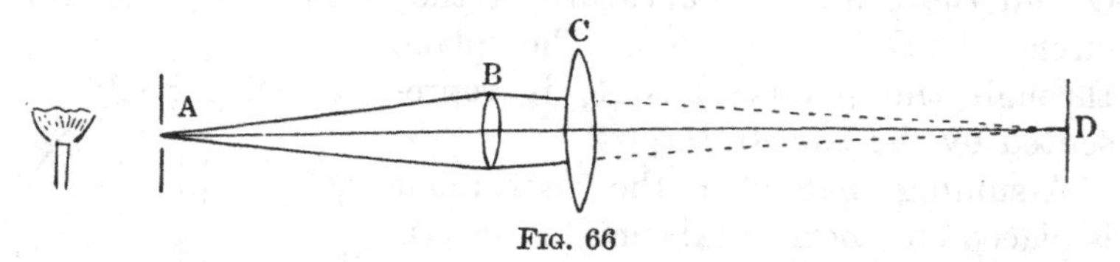

FIG. 66

stop of small diameter. A is the screen in the centre of which is a small hole with a cross wire, which is illuminated as shewn. Keeping the distance AB, which must be greater than the focal length of B, constant, we can find a position of C such that the rays of light after passing through B, strike the surface of C perpendicularly, and returning along the same path, form an image of the cross wire on A. Take the reading of the position of C, then remove C, and place a screen D in such a position that a distinct image of the cross wire is obtained on it, the lens B being kept in the same position. The distance from the lens to D is then the radius of curvature of the face nearer the cross wire. By turning the lens C round, the radius of curvature of the other face is similarly obtained.

If the lens B shews chromatic aberration appreciably substitute a sodium flame for the luminous flame.

If the focal length f of the lens is known the lens may be so placed that homogeneous rays falling on the second surface of radius s are normal to that surface and on reflection return to a focus at the cross wires u from the lens.

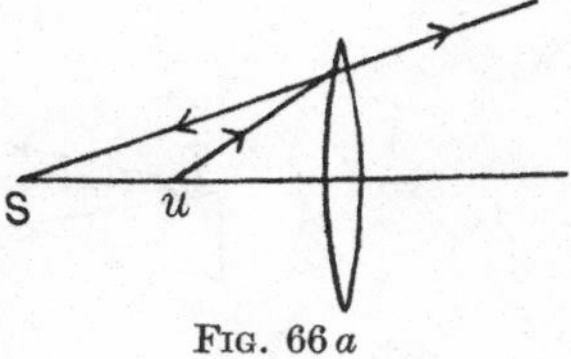

FIG. 66 a

Then
$$\frac{1}{s} = \frac{1}{u} - \frac{1}{f}.$$

The apparatus used is shewn in Fig. 67.

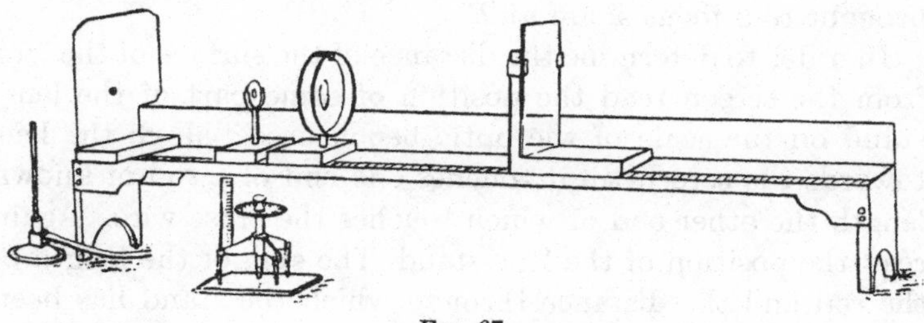
FIG. 67

Focal Length. Single Lenses.

Determine approximately the focal length of the lens, if convex, by placing a screen behind it at such a distance that an inverted image of some distant object is formed on the screen. The distance of the screen from the nearest point on the surface of the lens is the focal length approximately.

To determine it more accurately, if the lens is a converging one, mount it with a plane mirror immediately behind it. Place the mounted lens and the mirror on the optic bench.

Behind the hole and cross wire place a luminous burner, and find the position of the lens and mirror for which a distinct image of the cross wire is projected on to the screen, in the same plane and as nearly as possible coincident with the cross wires themselves (Fig. 68). Use as small a stop as possible between the lens and mirror, and if the lens shews chromatic aberration substitute a sodium flame for the luminous flame.

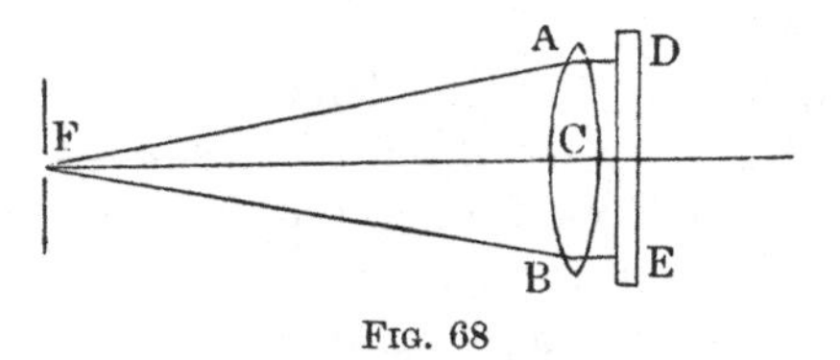

FIG. 68

The light from F in the direction FA is refracted by the lens and proceeds after refraction in the direction AD. If AD is perpendicular to the surface of the mirror, the reflected ray traverses the same path as the incident, and is, therefore, brought to a focus again at F.

In order to determine the distance of the surface of the lens from the screen read the position of some part of the lens-stand on the scale of the optic bench used. Move the lens towards the screen till it touches one end of a rod of known length the other end of which touches the cross wire. Again read the position of the lens-stand. The sum of the length of the rod and the distance through which the stand has been moved is the distance of the surface of the lens from the cross wire. The focal length of a symmetrical lens may be found from this by adding to it $\frac{1}{3}$ the thickness of the lens.

If the lens is a divergent one, place between it and the cross wires a convergent lens. Place behind the two and close to the divergent lens, a plane mirror (Fig. 69). Move the divergent lens and plane mirror till a distinct image of the cross

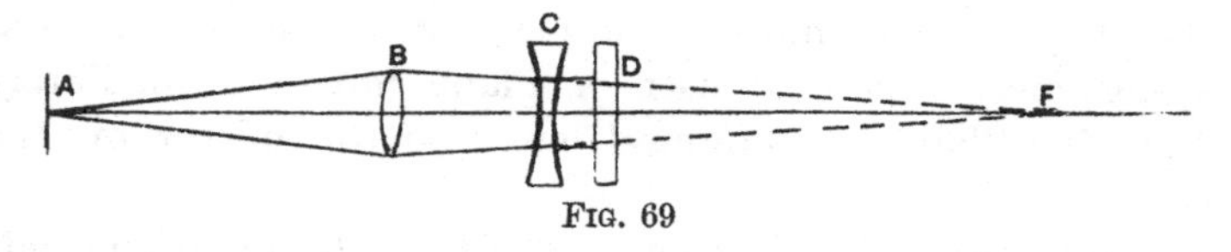

FIG. 69

wires is formed on the screen close to the wires. Note the position of the divergent lens, remove it and the mirror and place behind the converging lens a screen in such a position that an image of the cross wires is formed on it. The distance from the screen to the position previously occupied by the divergent lens is the focal length required. Use a stop between lens and mirror and a sodium flame if necessary.

The focal length of a lens in air is connected with the radii of curvature of its surfaces, its thickness d and with the index of refraction μ of its material for the light used, by the equation

$$\frac{1}{f} = (\mu - 1)\left(\frac{1}{R_1} - \frac{1}{R_2} + \frac{\mu - 1}{\mu}\frac{d}{R_1 R_2}\right),$$

or
$$\mu - 1 = \frac{1}{f}\Big/\left(\frac{1}{R_1} - \frac{1}{R_2} + \frac{\mu - 1}{\mu}\frac{d}{R_1 R_2}\right),$$

where R_1 and R_2 are taken positive for surfaces convex to the incident light.

The quantity $\frac{\mu - 1}{\mu} d$ is the distance apart of the principal planes of the lens.

Calculate μ from the observations, and tabulate as follows:

Date:

Lens No. 2 double convex, 1·02 cms. thick.

$a = 6{\cdot}14$ cms. $\quad \alpha = -{\cdot}03$ cm. $\quad \alpha^2 = {\cdot}0009$
$b = 6{\cdot}03$,, $\quad \beta = -{\cdot}14$,, $\quad \beta^2 = {\cdot}0196$
$c = 6{\cdot}33$,, $\quad \gamma = +{\cdot}16$,, $\quad \gamma^2 = {\cdot}0256$
$\therefore s = 9{\cdot}167$,, and $r = 3{\cdot}560$ cms. $\quad \therefore t^2 = {\cdot}023$
$\frac{2}{3}s = 6{\cdot}167$,, $\quad 9t^2/4s^2 = {\cdot}0006$,

$a' = 3{\cdot}55$ cms. $\quad \therefore r - a' = {\cdot}01$
$b' = 3{\cdot}54$,, $\quad r - b' = {\cdot}02 \quad \therefore d^2 = {\cdot}0005.$
$c' = 3{\cdot}57$,, $\quad r - c' = -{\cdot}01$,

By spherometer $h_1 = {\cdot}1564$ cm. $\quad h_2 = {\cdot}1567$ cm.
Hence $R_1 = 40{\cdot}60$ cms. $\quad R_2 = 40{\cdot}52$ cms.
By reflection $R_1 = 39{\cdot}9$,, $\quad R_2 = 40{\cdot}0$,,
f (measured) $= 40{\cdot}5$ cms.
Hence μ (calculated) $= 1{\cdot}50$,, for sodium lights.

When the power of the lens is small the images obtained in the previous methods are too weak to admit of accurate adjustments. The most convenient method is then that using an auxiliary telescope.

Determine the focal length F of the objective of the telescope by taking out the eyepiece, turning the eye end of the telescope towards a distant light and finding how far the image is from the objective $= F$.

Insert the eyepiece and focus the telescope on a distant object. Measure how much the eyepiece end has been racked out of the objective tube $= d_0$. Place the lens where focal length f is to be measured directly in front of the objective and refocus on the distant object. Measure again the distance the eyepiece has been racked out $= d_1$.

Then
$$f = \frac{F^2}{d_0 - d_1} - F.$$

Determine the focal lengths of the two spectacle lenses in this way, recording as follows:

Date:

Telescope B. Focal length of objective 16·3 cms.

Without lens	$d_0 = 4{\cdot}25$ cms.		
With lens 22	$d_1 = 2{\cdot}90$ „	$d_0 - d_1 = 1{\cdot}35$ cms.	$\therefore f_1 = 181$ cms.
„ „ 43	$d_1' = 5{\cdot}55$ „	$d_0 - d_1 = -1{\cdot}30$ „	$\therefore f_1' = -220$ „

Focal Lengths. Optical Systems.

For systems of small power whether converging or diverging the telescope method described above may be used.

For converging systems of higher power Cheshire's method is suitable.

A and A' (Fig. 69 a) are two divisions of a millimetre scale ruled on glass or celluloid. S is a slit 1 mm. wide in a thin metal sheet known as a telecentric stop, L the lens system whose focal length is to be determined, BB' a second millimetre scale and E a small lens or an eyepiece through which the scale BB' is viewed. The surfaces on which the scales are ruled are in each case turned towards the lens L. Place a strong light behind the scale AA' and a plane mirror

FIG. 69a

behind L, as in Fig. 68, and adjust S till a distinct image of itself appears by its side. S is then in the focal plane of L. Place AA' three or four times as far from the lens as S is and adjust BB' till AA' appears in focus on it. Observe the ratio BB'/AA', and measure the distance a between AA' and S. Then $f = a\,\dfrac{BB'}{AA'}$. Check by reversing the lens system.

Record as follows:

Date:

Lens system C. Wide end towards stop.

Distance of telecentric stop from middle of lens mount... ... ... = 12·52 cms.

Distance of scale behind stop... ... ... = 30·00 ,,

$AA' = 2$ cms. $BB' = \cdot 73$ cm. $BB'/AA' = \cdot 365$ cm.

$\therefore f = 10\cdot 9$ cms.

Similarly with the narrow end towards stop.

SECTION XXXIV

INDEX OF REFRACTION BY TOTAL REFLECTION

Apparatus required: *Horizontal graduated scale with upright and slits, ebonite block, and half cylinder of glass.*

When a ray of light traversing an optically dense medium impinges on the surface of separation of that medium from a rarer medium, making an angle with the normal at the point of incidence greater than the "Critical Angle," the ray is totally reflected, no part of it entering the rarer medium. If N is the index of refraction of the denser, n that of the rarer medium, the least value of the angle of incidence for which total reflection can take place, *i.e.*, the Critical Angle, is given by the equation $\sin \theta = n/N$.

A half cylinder of glass of about 2 cms. edge is provided. On the plane face the axis of the cylinder has been drawn with a diamond. Place the half cylinder face downwards on the ebonite block provided, and the ebonite on the scale about 20 cms. behind the slit in the upright, with the axis parallel to the slit. Place the scale on a table in front of a window through which the sky can be seen. Look through the slit at the lower surface of the cylinder, and notice that on moving the cylinder and block towards the slit, the appearance of this surface changes from bright to dull, the bluish line of change extending across the surface parallel to the axis, the nearer part of the surface appearing the brighter. Adjust the distance of the cylinder from the slit till the line of demarcation coincides with the diamond scratch at the axis.

Measure h the height of the slit above the surface of the scale, and d the distance along the scale between the foot of a perpendicular let fall on to the scale from the slit, and the apparent position on the scale of the axis of the cylinder.

Turn the block and cylinder 180° about a vertical axis and repeat the observation of d. Table the mean of the two.

Put a drop of water on the block so that when the cylinder is placed on it the plane surface of the cylinder is in contact with water. Repeat the observation of d using the lower slit and measuring h'.

Dry block and cylinder, substitute alcohol for water and repeat.

Then if θ = angle of incidence of the last ray reflected from the surface (Fig. 71), $\tan\theta = h/d$.

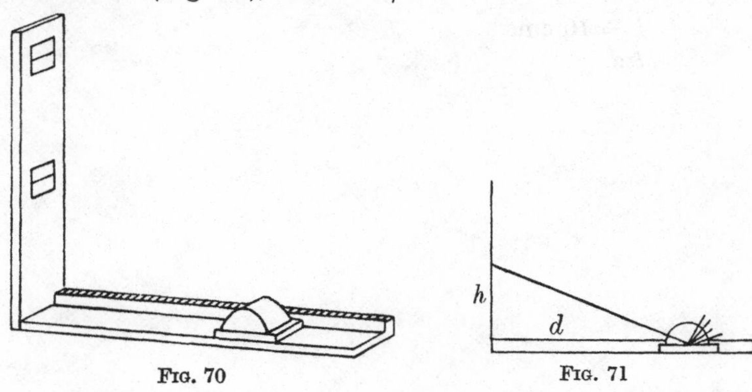

FIG. 70 FIG. 71

If N = the index of refraction of the glass, n that of the drop for the light used, or for the middle of the spectrum near F if the light is white, we have

$$n = N\sin\theta \text{ or } N = n\sqrt{1+\frac{h^2}{d^2}}.$$

From which N or n may be calculated if one is known and θ is found by measurement of h and d.

Determine by this method the value of N for the glass for the line F, taking n for air to be 1·0.

Having found N find n for water and ethyl alcohol.

Arrange your observations as follows:

Date:

Air film h = 20 cms.

d = 15·6, 15·7, reversed 15·5, 15·6, mean 15·6 cms.

$$h/d = 1{\cdot}28,\ h^2/d^2 = 1{\cdot}64,\ \sqrt{1+\frac{h^2}{d^2}} = 1{\cdot}62.$$

$$\therefore N \text{ for glass} = 1{\cdot}62.$$

Water film

$$h = 10 \text{ cms.}$$

$$d = 14{\cdot}8,\ 14{\cdot}7,\ \text{reversed } 14{\cdot}7,\ 14{\cdot}6,\ \text{mean } 14{\cdot}7 \text{ cms.}$$

$$h/d = {\cdot}681,\ h^2/d^2 = {\cdot}464,\ \sqrt{1 + \frac{h^2}{d^2}} = 1{\cdot}21.$$

$$\therefore n \text{ for water} = \frac{1{\cdot}62}{1{\cdot}21} = 1{\cdot}34.$$

Ethyl alcohol film

$$h = 10 \text{ cms.}$$

&c.

SECTION XXXV

RESOLVING POWER OF A LENS

Apparatus required: *Good lens (a telescope objective), observing eyepiece, two slits or fine wire gauze, sodium flame and stops.*

If a cylindrical lens AB (Fig. 71 *a*) forms an image Q of an infinitely narrow luminous line P parallel to its axis, that image is a narrow strip of light whose width is determined by the fact that the difference of the distance of either edge of the strip from A and B is λ the wave length in the medium about the lens of the light used.

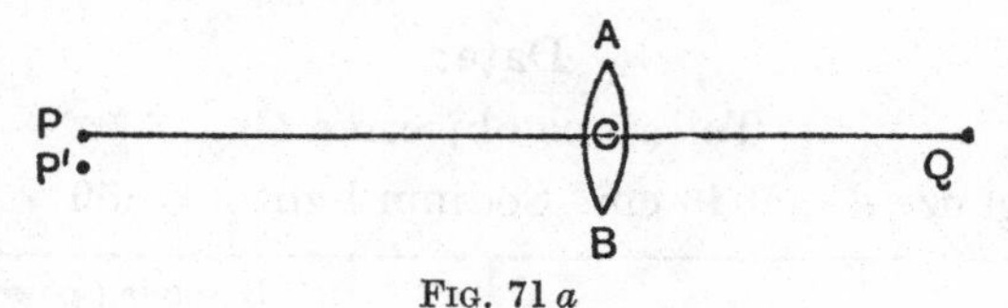

Fig. 71 *a*

It is found that another infinitely narrow line P' at a distance δ from P produces a strip image it is just possible to distinguish from the first if its edge coincides with the centre of the first image, that is if

$$P'A - P'B = \lambda.$$

Hence the images of two lines δ apart would just be separated or "resolved" if

$$\delta/CP = \lambda/AB,$$

where δ/CP is the angular distance of the two lines apart or angular separation of the lines, as seen from the centre of the lens. Hence a cylindrical lens resolves two close line sources if their angular separation $= \lambda/AB$.

The resolving power of the lens is defined as the reciprocal of the least angular separation of two lines just resolvable and is therefore $= AB/\lambda$.

If the lens is circular instead of cylindrical its effective width is ·8 of its diameter and its resolving power $\cdot 8AB/\lambda$*.

Set up the mirror with two fine parallel slits in the silvering or the wire gauze, with the slits vertical, place a sodium flame behind it, and observe it through a telescope from a distance D. On focussing on the screen the slits or spaces between the wires of it may be seen distinctly if D is not large. Increase D, focussing each time till the separate slits or spaces are no longer distinguishable. Measure D and count how many wires of the grating there are in 1 cm. breadth. The reciprocal of this is d the distance from centre to centre of the spaces and D/d the observed resolving power of the telescope.

Place the screen with the circular hole 1 cm. diameter over the objective of the telescope and repeat the observations. Do the same with each of the 5 screens provided, measuring the breadth of the hole in each case. Tabulate the results as follows:

Date:

Telescope objective C.

Wire gauze $d = \cdot 046$ cm. Sodium light $\lambda = \cdot 59 \times 10^{-4}$ cm.

Hole	Breadth AB	D	Resolving power	
			observed $=D/d$	theoretical AB/λ or $\cdot 8AB/\lambda$
Objective	2·5 cms.	1900 cms.	41300	34000
Circular	1·0 cm.	760	16500	13500
,,	·5	375	8500	6800
,,	·25	180	3900	3400
Slit	·43	505	11000	7300
,,	·25	270	5850	4100

It will be noticed that the observed resolving power exceeds the theoretical, which is generally found to be the case in good lenses.

* See Rayleigh, *Encyclopaedia Britannica*, article "Diffraction of Light," § 5, or Schuster, *Optics*, chap. VII.

SECTION XXXVI

THE PRISM SPECTROSCOPE

Apparatus required: *Prism spectroscope, platinum wires for beads, salts, and crayons.*

Vision through a prism. Let a luminous point S (Fig. 72) send out a pencil of homogeneous light, the rays of which, after refraction through the prism, seem to diverge approximately from a point S'. An eye at E will therefore see the luminous point in a displaced position, the amount of displacement depending on the refractive index of the prism for the rays and on its angle. If the light sent out by the luminous point is not homogeneous, but is of two colours, for which the prism has different refractive indices, there will be two separate coloured images side by side, if the difference in the refractive index is sufficiently great. If, finally, the luminous point sends out white light, there will be an infinite number of images shewing a succession of all the colours of the spectrum from red to violet. If the point S is replaced by a luminous slit parallel to the refracting edge of the prism the images will be lines parallel to the slit.

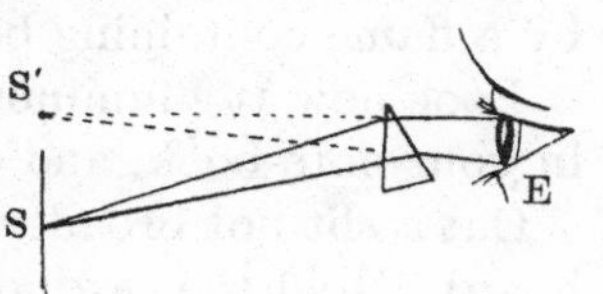

Fig. 72

Seal about four or five centimetres of platinum wire of ·2 mm. diameter into a piece of glass tubing, and bend the end of the wire into the shape of a loop (Fig. 73). Wet the loop slightly, dip it into a mixture of common salt and borax in equal proportions, and heat carefully in the edge of the flame of a Bunsen burner, until the salt fuses and forms a transparent bead. If the loop is not completely filled by the bead, place some more of the substances on it and repeat

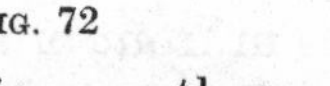

Fig. 73

the process of fusion. Then fix the glass tube to the stand provided, and place it so that the bead touches the flame and colours it. The platinum wire should dip slightly downwards (Fig. 73), otherwise the bead has a tendency to run back along it.

Hold the loose prism provided in front of your eye, and turn your head until you see the displaced image of the flame. Next take a second piece of wire, and having prepared a bead of a lithium salt, place it in the flame along with the sodium bead. Make a sketch in your note-book of what you see. What conclusions do you draw respecting the light sent out by a flame containing both sodium and lithium?

Look now at a luminous flame, again sketch the appearance in your note-book, and explain it.

Cut a slit not broader than a millimetre in a piece of cardboard. Hold the cardboard in your left hand so that the slit is in front of a Bunsen burner containing both a lithium and sodium salt, hold the prism in the right hand in front of your eye and look at the image of the slit. Sketch the appearance in your note-book.

Such a combination of slit and prism forms the simplest kind of spectroscope, and is often useful for the rapid examination of light sent out by a flame, or an electric discharge.

In practice the virtual image of the slit produced by the prism is magnified by being looked at through a telescope. But this cannot be done with advantage without some further change in the optical arrangement, to obviate the so-called "aberration" of the rays. For if rays of light diverging from a point are traced through a prism, it is found that after emergence they do not accurately diverge from a point, but the section of the pencil will have the appearance of Fig. 74. This does not sensibly affect the sharpness of the image when looked at with the naked eye, or even under

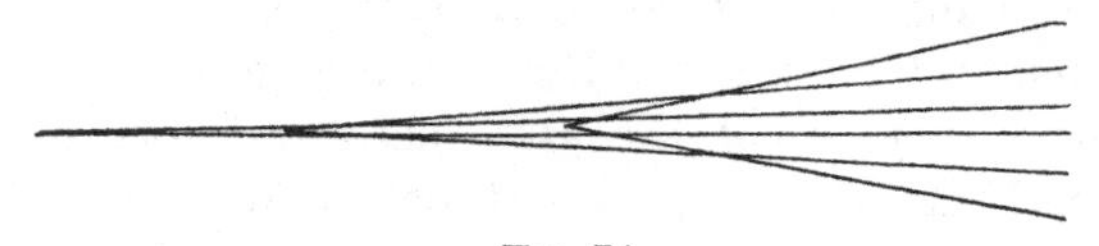

Fig. 74

small enlargements such as are used in pocket spectroscopes, but when higher power is required the aberration must be obviated. This may be done by the introduction of a lens placed in such a position between the slit and the prism that the beam of light coming from any point of the slit becomes parallel before it falls on the prism. As a parallel pencil of light will remain parallel after refraction at any number of plane surfaces, there is no aberration in this case. We thus arrive at the arrangement shewn in Fig. 75. A tube carries

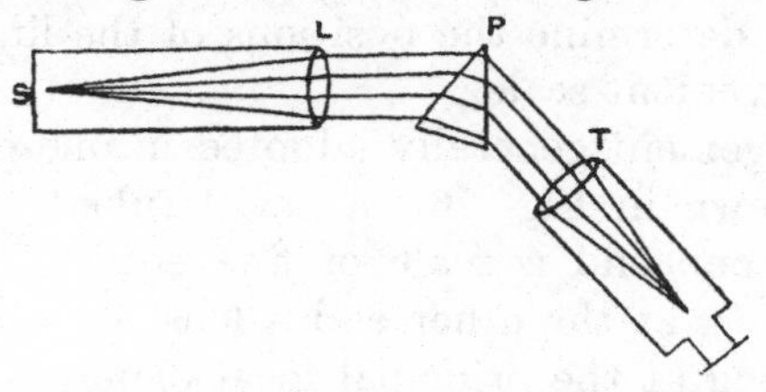

FIG. 75

at one end the slit S and at the other the lens L, the slit being in the focal plane of the lens. This tube is called a "collimator." The pencils of light from the slit rendered parallel by L fall on the prism P, and after passing through it are received by a telescope T. The eye, on looking through the telescope will see an image of the slit, which is displaced by the prism, but is sharp provided the slit is illuminated by homogeneous light. If, on the other hand, the light falling on the slit consists of groups of waves differing in wave length, each group will give a separate image of the slit. If further the wave lengths, and therefore the refrangibilities, vary continuously over a certain range, the images of the slit will lie side by side or even overlap, so that a continuous band of light will be seen.

We call the appearance presented when the light of a luminous body is examined by means of a spectroscope the "spectrum" of the body. We say that the body has a "continuous spectrum" when the band of coloured light is continuous. We say, on the other hand, that the body has a "line spectrum," if a number of separate coloured lines are shewn, which we have seen are only images of the slit. If the slit is curved the lines will be curved, if the slit is broad the lines appear broad, and narrowing the slit will narrow the lines down to a certain limit, which depends on the diameter of the object-glass used. A "band spectrum" is a spectrum consisting of bands which are broad even with a narrow slit. These bands are often sharp on one side and fade away gradually on the other.

As the spectrum of a body, whether it is a line or band spectrum, is found to be characteristic of the body, it is necessary to determine the positions of the lines and bands on some convenient scale.

The arrangement generally adopted in one-prism spectroscopes is shewn in Fig. 76. A small tube MQ has at one end a scale of fine equidistant lines Q, at the other end a lens M, the scale being at the principal focal plane of the lens. The tube is placed so that the light rendered parallel by the lens M is reflected at the surface of the prism into the telescope T. When the scale is illuminated by a small gas flame, the observer sees not only the spectrum of the body, but superposed on it the image of the divided scale, and he can read off the position of each separate line on this scale.

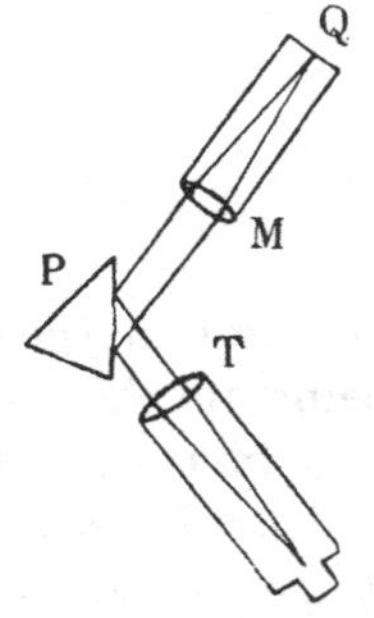

FIG. 76

Fig. 77 shews a spectroscope consisting of collimator, prism, telescope, and scale tube. The collimator is provided with a projecting metal sheet, and the scale tube with a projecting metal cylinder, to prevent the flames used being brought so near as to injure the instrument.

Adjustments of the Spectroscope: It has been shewn above that the light leaving the collimator should consist of parallel rays. To secure this, if the distance between the slit and collimator lens can be altered, the telescope and collimator are adjusted as follows:

Place the spectroscope near a window and turn the telescope till a distant object is observed through it. If the object is seen clearly, the telescope is in adjustment for parallel light.

If not adjust by means of the focussing screw. Remove the instrument to a darkened room, illuminate the slit of the collimator by a sodium flame, and move the collimator in and out till the image of the slit seen in the telescope is sharp and distinct. Place a small luminous burner in front of, but *not too near* to, the end of the scale tube, and adjust the scale tube till both the scale and the sodium line are seen sharp at the same time. Move the scale tube at the same time until the sodium line stands at a convenient number of the scale*. The best test of good adjustment is the absence of

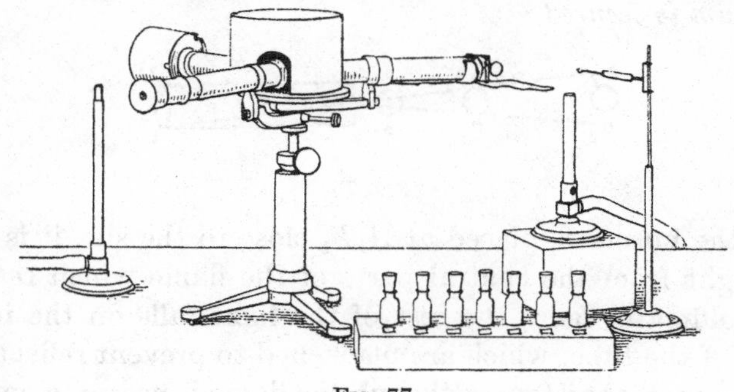

FIG. 77

parallax; that is to say when the eye is moved a little to the left or right the sodium line should not change its position relative to the scale. Accurate measurements with the spectroscope are impossible when parallax exists to an appreciable degree.

If the position of the slit of the collimator cannot be changed, the telescope must be adjusted by illuminating the slit with a sodium flame, and focussing the telescope so that the yellow line appears sharp. The scale is then focussed as described.

Before proceeding with any further measurements, consult one of the Demonstrators to see that your adjustment is correct. With a narrow slit the sodium line should just appear double.

As it is generally necessary to have the spectra as bright

* As the scale divisions may differ in different instruments, a label is attached to each spectroscope, on which the number with which the sodium line should be made to coincide is indicated.

as possible it will be useful to remember the following facts. Let SL (Fig. 78) be a horizontal section of the collimator, S being the slit and LL' the lens. Let AB be a flame sending out light the spectrum of which is to be examined. The light from the part of the flame near A will illuminate the part of the collimator lens near L, the rays of light going from A through the slit to L; similarly the light from the part of the flame near B will illuminate the part of the lens near L'. The whole collimator lens is thus filled with light. *When the whole collimator lens is filled with light maximum brightness of the spectrum is secured.*

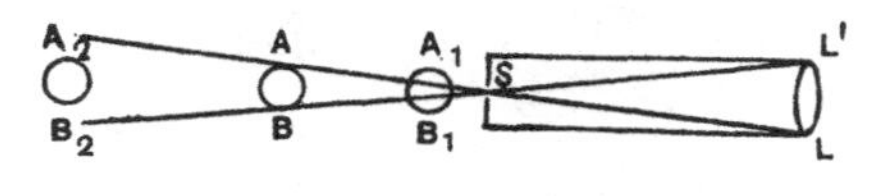

FIG. 78

If the burner is placed at A_1B_1 close to the slit, it is only the light from the central parts of the flame which reaches the collimator lens; the rest of the light falls on the inside walls of the tube, which are blackened to prevent reflections. In this case, therefore, although the flame is nearer, a smaller part of the flame illuminates the collimator lens, so that nothing is gained in brilliancy, the brightness of the spectrum being exactly the same as before. If, on the other hand, the flame is too far away at A_2B_2, the outer portions of the lens are not illuminated at all; in that case the brightness of the spectrum is reduced. When the collimator lens is not filled with light, as in this last case, beside loss of light there is the disadvantage of worse definition, so that in spectroscopic work it is generally of importance that the whole collimator lens should be illuminated, but when that is secured no further increase in brightness is possible.

Where the source of light is very small, as in the case of an electric spark, with a narrow slit the collimator lens would not be filled with light unless the source were brought inconveniently near the slit. A lens P (Fig. 79), should therefore be placed in such a position between the spark E and the slit S that the rays traverse the paths shewn, and an image of the spark is formed on the slit. Here, again, it does not matter how the distances of lens and spark from the slit are adjusted, nor how bright is the image of the spark seen on the slit plate. The filling of the collimator lens with light is the sole circumstance on which the brightness of the spectrum depends. We may even move the lens so that the image of the slit is out of focus without producing great loss of light. The use of the lens has the further advantage of enabling us to examine the spectra of different *parts* of the spark separately.

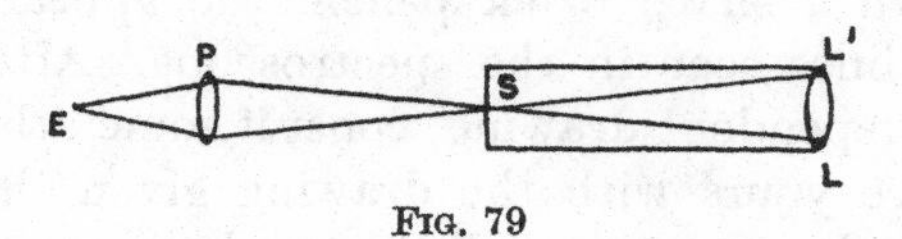

FIG. 79

Mapping Spectra: When the spectroscope is in adjustment and the sodium line at the proper scale division, place a luminous burner behind the slit of the collimator so as to produce a continuous spectrum. Draw the spectrum which you see, representing as nearly as possible with the crayons provided the colours and shades and their extent on the spectroscope scale. Use for this purpose a sheet of paper on which scales with the distance between the scale divisions approximately equal to one millimetre are printed.

Next prepare six pieces of platinum wire sealed into glass tubes, as described at the beginning of the section. The platinum should not be thicker than ·2 mm.; it must be perfectly clean, and before use be heated to a red heat in the Bunsen flame.

Adjust the height of the burner so that the hottest part of the flame, which is just above the top of the inner cone, lies about the level of the lower portion of the slit.

Place a small quantity of barium chloride on the platinum wire, wetting the latter first with a little dilute hydrochloric acid. Support the wire in the stand provided and insert it into the hottest part of the flame. After a few seconds the upper part of the flame will be green coloured, and on looking through the spectroscope you should see a spectrum of bands, with a line in the green which is sharp and bright. If this line does not appear, the adjustment of your spectroscope is probably faulty or your slit is too wide. If the spectrum is very faint, either the flame is not placed properly in front of the slit, or the latter is too narrow. The best width of slit is only obtained by trial. For the observation of the spectra of the alkali metals with small spectroscopes, the slit ought not to be so narrow that the sodium line appears double, but it ought to appear a sharp line and not a broad band.

Represent on the paper scales provided, by means of shading with a sharp black pencil, the appearance of the bands and lines seen in the spectroscope. After you have made an independent drawing, consult some atlas of spectra and compare yours with the drawing given there. If you find that you have not seen all the bands shown there, repeat your experiment. Some of the fainter bands may however escape your notice, as they require a completely darkened room.

Make similar drawings of the spectra of lithium, thallium, strontium, and calcium chlorides. Notice that the spectrum of the last-mentioned salt gradually changes. The spectrum first seen is due to the chloride of calcium, and is gradually replaced by that of the oxide. Make separate drawings of the first and last stages through which the spectrum passes.

During the observations, check from time to time the position of the sodium line. If you find that a slight change has taken place, it is not necessary to alter the position of the scale tube, apply instead a correction to your readings. Thus, if the sodium line reads 8·1 instead of 8, you may correct all other readings by subtracting ·1.

As a last example take a bead of potassium chloride. You should have no difficulty in seeing a line in the red, which is really a double line. But there is also a line at the extreme end of the visible violet, which requires special precaution to see. Not to have your eye disturbed by extraneous light, move the telescope until the red and green are out of the field of view, so that when looking at the continuous spectrum of a gas flame you only see the violet; also remove the burner which illuminates the scale. Carefully introduce the potassium salt into the hottest part of the flame while you are looking through the telescope; with a little practice this is easily done. If you fail to see the potassium line, slightly open the slit. As soon as you can see it, illuminate the scale a little and take a reading. Finally, move the telescope back, and without altering the width of the slit, take a reading of the sodium line. The measurement of the potassium lines should be made several times in order to secure greater accuracy.

It will be noticed that while thallium, sodium, potassium, and lithium give line spectra, the spectra of the alkaline earths consist of bands. These latter are due to the metallic oxides or chlorides, while in the case of the alkali metals the salts are decomposed and the spectra of the metals themselves appear.

Instead of making shaded drawings of the appearance of spectra, they may be represented diagrammatically, as in Fig. 80, where the intensities are represented by the lengths of the lines. Thus *B* is a sharp line, the intensity of which is estimated to be about $\frac{1}{3}$ that of another sharp line *A*, *D* a band having a sharp edge at 11·30 but fading away gradually to the other side, the limit of visibility being about 11·7, *C*, on the other hand, a band which is brightest in the middle and ends abruptly at 13·6 and 13·84.

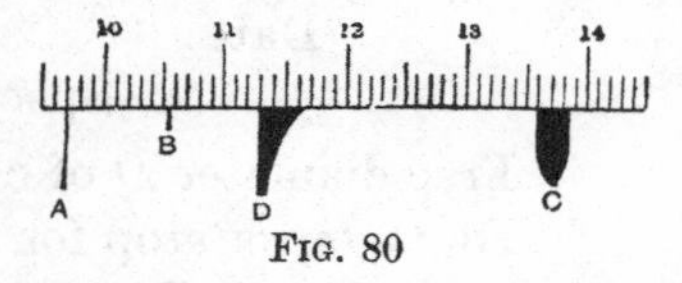

FIG. 80

Resolving Power*.

The resolving power of a spectroscope for a given part of the spectrum is defined as the quotient of the wave length of the line of light used, by the least change of wave length which in the instrument shews as a line distinct from the former.

It is shewn in text-books to be equal, when a fine slit is used, to the product of the breadth of the beam of light entering the telescope by the rate of change of the deviation produced by the prism with the wave length of light passing

* Kayser, *Spectroskopie*, Vols. v and vi. Eder and Valenta, *Atlas Typischer Spectren.*

through it. By reducing the width of the beam we may therefore diminish the resolving power.

To do this place stops of diameters varying from 2 to $\frac{1}{2}$ cm. in succession before the objective of the telescope. Note that when the width of the beam of light entering the telescope is reduced to ·5–1·0 cm., depending on the material of the prism, the two sodium lines cease to be distinguishable from each other; that is, they are no longer resolved. Measure the diameter d of the hole in the last stop with which the line is resolved and the free diameter D of the objective of the telescope. When the sodium lines $\lambda = 5890{\cdot}0$ and $5895{\cdot}9$ are just resolved the resolving power of the instrument is by definition $\frac{5890}{5{\cdot}9} = 1000$. Hence when the whole width of the objective is used by the beam from the prism the resolving power is $1000D/d$.

Record as follows:

Date:

Spectroscope B.

Free diameter D of objective ... 2·2 cms.
Diameter of stop for resolution ... 1 cm.

$\therefore$ Resolving power = 2200.

SECTION XXXVII

REDUCTION OF SPECTROSCOPIC MEASUREMENTS TO AN ABSOLUTE SCALE

WHEN the spectrum of a given substance is mapped, the relative positions of the lines will be found to differ according to the instrument used. Even if the sodium line is at the same scale division in each, the readings of the other lines will differ in the different instruments, since they depend on the dispersion of the prism and on the distance between the divisions of the scale. In order that measurements taken with different instruments may be comparable, it is necessary to have some method of representation which is independent of the instrument employed.

It is usual to take the wave length in air as that which characterises a given light vibration, and the spectrum mapped according to the wave length scale is called a "normal" spectrum. Instead of the wave length scale, we might adopt the "wave number," *i.e.* the number of waves per centimetre, or the "frequency" as the characteristic property of a set of waves, and such scales present many advantages. The frequency n is connected with the wave length λ by the relation $n\lambda = v$, where v is the velocity of light. If a set of waves in air, from a source sending out homogeneous light, enters a different medium, as for instance a glass prism, the frequency will be the same in the glass as in air, but the velocity of propagation of the waves, and hence the wave length, will be different. When a particular kind of homogeneous light is defined by its wave length, we must therefore state the medium in which that length is measured.

The most scientific proceeding would be to take the wave length *in vacuo* as the standard, but as all direct measurements of wave lengths are taken in air, and the reduction to vacuo involves an accurate knowledge of the refractive index of air, it is usual to take the wave lengths in air as standards. To be strictly accurate we must define the pressure and the temperature of the air, since its refractive index depends on its density. The table of standard wave lengths of the International Astronomical Union is generally used, and holds for a pressure of 76 cms. and a temperature of 15° C. In certain investigations it is necessary to know the wave lengths *in vacuo*, and Tables have been constructed (Watts's *Index of Spectra*) by means of which the conversion to vacuo may be easily effected.

In order to avoid having the decimal point at an inconvenient place, wave lengths are usually given not in centimetres but in the 10^8th part of a centimetre known as an Ångström unit and written Å. According to this scale the wave lengths of the two sodium lines in air are 5890·18 and 5896·15, and the correction to vacuo is + 1·60. Spectroscopes containing one prism of the usual size only barely separate the sodium lines, and measurements with such instruments can only be accurate to two or three Å units. All figures beyond the decimal point may therefore be left out of account in what follows.

The laws of spectral series are most conveniently expressed in terms of wave numbers, or *number of waves per centimetre* rather than the wave lengths. If there are ν waves in one centimetre $\nu\lambda = 1$, and hence the wave number is simply the reciprocal of the wave length. For the least refrangible sodium line the wave number is therefore

$$\frac{10^8}{5896 \cdot 15} = 16960 \cdot 2.$$

Different methods may be used to reduce measurements taken on the instrument scale to wave lengths. In each method the positions of two or more lines, the wave lengths of which are known, are observed on the scale of the spectroscope, and the unknown wave lengths of other lines obtained from them by interpolation. It is in the process of interpolation that the methods differ.

It is simplest to draw a curve connecting the scale readings with the wave lengths. To do this, the positions of a certain number of so-called "standard lines," p. 191, are observed on the spectroscope scale, and a curve having the scale divisions as abscissae and the wave lengths as ordinates is drawn. The shape of the curve will resemble that shown in Fig. 81.

Its curvature is too great for accurate work. On the left in the region of greater wave lengths a small error of reading makes a large error in the wave length, on the right the opposite holds.

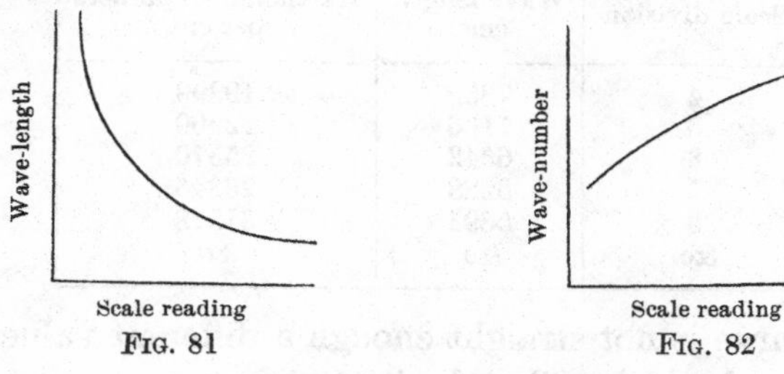

FIG. 81 FIG. 82

To obtain a curve which is more nearly straight we make use of the wave numbers ν as ordinates instead of wave lengths λ and obtain a curve like that of Fig. 82. The curvature is still greater than is desirable and an error of reading makes a greater error for the smaller than for the greater wave lengths.

Exercise I. Draw a curve connecting the inverse wave lengths or wave numbers with the scale readings. Prepare a table from this curve, thus:

SPECTROSCOPE A. *Sodium lines at* 8.

Scale division	Wave-length cms.	Wave numbers per cm.
4	7860×10^{-8}	12723
5	7238	13816
6	6712	14895
7	6277	15931
8	5893	16969
&c.	&c.	&c.

Hartmann's method consists in plotting $1/(\lambda - \lambda_0)$, where $\lambda_0 = 2700 \times 10^{-8}$ cms. approximately, against the scale reading.

It gives a curve very nearly straight, but it involves an increased amount of calculation.

Exercise II. Draw a curve connecting Hartmann's numbers with the scale readings. Prepare a table from the curve as follows:

Date:

SPECTROSCOPE A. *Sodium line at* 8.

λ_0 taken 2700×10^{-8} cms.

Scale division	Wave length cms.	Hartmann's numbers per cm.
4	7855	19399
5	7166	22390
6	6542	25370
7	6228	28345
8	5893	31318
&c.	&c.	&c.

If the curve is not straight enough a different value for λ_0 may be found which will make it straighter.

When a value of λ_0 has been found which gives a straight line interpolation between the standard wave lengths it may be done arithmetically with greater accuracy than by the graphical method.

Calling y_1, y_2 and y the Hartmann numbers for the two reference lines and for the unknown line respectively and n_1, n_2, n the scale readings, we have:

$$y = y_1 + \frac{n - n_1}{n_2 - n_1}(y_2 - y_1).$$

From y, the wave length is found by calculating $\lambda_0 + 1/y$.

For more accurate work a table of "divided differences" is constructed. See p. 9 and references there.

Exercise III. Map a spectrum on the normal scale. Using the curve obtained in Exercise II, make as accurate a drawing as you can of the calcium spectrum on a scale of wave lengths, representing the characteristic features of the spectrum by one of the two methods of mapping spectra described on pages 183–185.

Exercise IV. Using the measurements previously made, calculate, by the graphical and arithmetical methods above, the wave length of the sodium line, having given the wave lengths of the lithium and thallium lines.

Enter observations and results as follows:

Line	Scale division	Wave lengths in Å units		
		Given	Calculated graphical	Calculated arithmetical
Lithium	6·15	6708		
Thallium	10·55	5351		
Sodium	8·0		5932	5898

Correct Wave length = 5893 Å.

Notice that the graphical method gives only an approximate value, but that the arithmetical method gives a fairly accurate result. For very accurate work the reference lines should be nearer to the line to be determined.

TABLE

Reference lines	Wave length X^{th} Metres in air	Wave number per Centimetre in air
K red. Centre of double line	7682·5	13016·6
Li red	6708·2	14907·1
H_α red (C)	6563·05	15236·8
Na yellow. Centre of double line (D)...	5893·17	16968·8
Tl green	5350·65	18689·3
H_β green (F)	4861·50	20569·8
Sr blue	4607·51	21703·6
Ca blue	4226·90	23658·0
K violet. Centre of double line ...	4045·35	24719·7

The wave lengths and wave numbers are here given to six significant figures, as the more complete values may be useful for reference, but for the purposes of the above exercises it will be sufficient to use four figures. The wave lengths are taken from the table of the International Astronomical Union.

SECTION XXXVIII

THE SPECTROMETER

Apparatus required: *Spectrometer, with Gauss eyepiece, plane parallel glass on stand.*

An instrument in which the deviation produced by the passage of a beam of parallel light through a prism or other apparatus can be measured is called a spectrometer. It consists of a collimator S, a telescope T, and a horizontal divided circle C (Fig. 83). The telescope moves about a vertical axis passing through the centre of the circle. Attached to the telescope is an arm carrying two verniers V, V' opposite

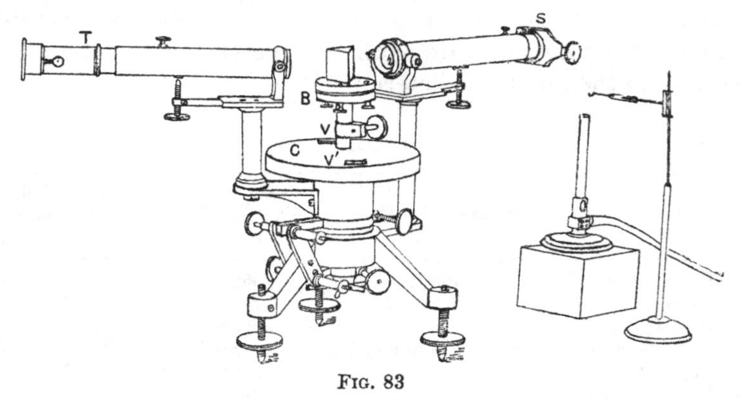

Fig. 83

each other, which move along the circular scale, so that the angle through which the telescope is turned may be accurately measured*. Above the divided scale is a table B, on which prisms or gratings may be placed. In some instruments this table is fixed, but it ought always to be capable of being turned about an axis coincident with the axis of rotation of the telescope. The way in which an angular displacement of

* In Fig. 83 the circle and verniers are enclosed in a case, and are read through windows in the case.

the table is measured differs in different instruments. If the divided circle is fixed to the collimator, the table should carry a second set of verniers. But in many instruments the divided circle is attached to the table. In that case, care must be taken that between any two readings of the verniers either the table only, or the telescope only is moved. If both are displaced, the vernier readings will only shew the difference between the angular displacements of table and telescope, not the actual displacement of either.

It is difficult to secure that the axis of rotation passes accurately through the centre of the divided circle. The error thereby introduced, called "error of eccentricity," is eliminated by having two verniers opposite each other as described. The angle measured by means of one vernier exceeds the correct value by as much as the angle measured on the other falls short of it; so that the arithmetic mean of the two results gives the correct angle. For a proof of this see Note, p. 200.

Examine the scale and vernier, and determine the value of the smallest subdivision of the principal and vernier scales. If *e.g.* the circular scale is subdivided into 20 minutes of arc, the vernier will probably be divided into 20 parts, of which every fifth will be numbered, and each of the 20 may be subdivided into 2 or 3 parts, which will allow the measurements to be made to 30 or 20 seconds of arc. In entering an observation in your note-book, write down separately the readings of the two scales. Thus if the angle to be read off were 47° 43′ 20″, the observation would be entered as follows:

Principal Scale	47° 40′ 0″
Vernier	3′ 20″
	47° 43′ 20″

A wooden model of the vernier is provided in the laboratory, and the student should practise reading it until he is quite familiar with it.

Before proceeding with the exercises students should study the construction of the instrument they are using, and refer to a more detailed description, which will be supplied with the instrument. Special care should be taken to be familiar with the object of the various screw heads, which serve either to clamp some part of the instrument, or to give that part a slow motion. If the telescope, for instance, is to be pointed in any direction, it is first moved by hand as nearly as possible into its right position. It is then "clamped" by the proper screw, and finally brought to the proper position by means of a "fine adjustment screw," which can, within certain limits, alter its direction. *For the purpose of clamping, it is not necessary to use great force. If the screw is screwed up* **gently** *it will be sufficient. If force is used the instrument will be damaged.*

Other moveable parts of the instrument will also in general be provided with a clamping arrangement and fine adjustment.

Before moving any part of the instrument care must be taken that the clamping screw of that part is released. All parts should move easily, and no force should ever be used.

In order to fix the direction in which the telescope points, a mark is placed in the focal plane of the object lens. This mark consists generally of a cross formed by two cocoon fibres, spider's threads, or very thin platinum wires. This arrangement, called the "cross wires," can be turned in its own plane, so that one of the wires may be placed vertically, or they may both be placed at an equal inclination to the vertical. The latter position is most convenient when the telescope is to be pointed towards the image of a slit, for it is found easier to place the centre of the cross on the image

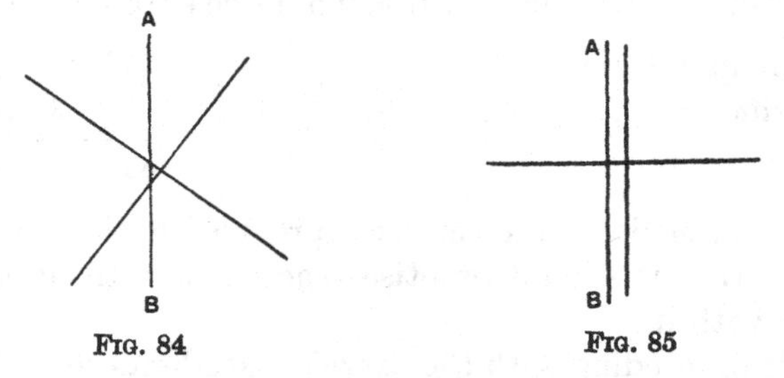

FIG. 84 FIG. 85

when the cross has the position shewn in Fig. 84, where AB represents the image of the slit, than when one of the wires is vertical, as in Fig. 85. For when the slit is narrow, as it should be, the vertical wire will be difficult to see against the dark background.

When the field of view is dark, it may be necessary to throw some light into the telescope in order to see the cross wires distinctly. This is best done by means of a sheet of white paper held in the hand and placed obliquely near the object-glass of the telescope; the light scattered from the white paper will illuminate the field of view sufficiently.

To focus the eyepiece on the cross wires, there must be some means of altering the positions of one of the two. If the cross wires are fixed in the draw tube, the eyepiece may be moved slightly outwards or inwards, if the eyepiece is fixed the wires are mounted on a frame which admits of a small motion. Turning the telescope towards a bright surface, such as an illuminated sheet of paper, move the eyepiece or the cross wires until you see them distinctly. The eye will be least fatigued if the cross wires are so placed that their image is as far from the eye as is possible, consistently with distinct vision.

To focus the telescope turn it towards some distant object, and alter the distance between object lens and eyepiece till a distinct image of the object is produced at the cross wires. Move the head sideways to see that there is no parallax.

If a parallel beam enters the telescope in such a direction that it converges to the central point of the cross wires, a line drawn through that central point parallel to the original direction of the beam is the "line of sight" of the telescope. If the light comes from a star, the line of sight is the line drawn from the centre of the cross towards the star. This line of sight should be perpendicular to the axis of rotation of the telescope (see p. 197).

In order to adjust the collimator to give a parallel beam of light, turn the focussed telescope so as to look straight into the collimator tube. Place a luminous burner 10 cms. behind the slit of the latter and focus the collimator until the edges of the slit look perfectly sharp. Place the cross wires on it, move the eye to the right or left and see that there is no parallax.

The level of the collimator must now be fixed, so that a point near the centre of the slit shall have its image covered by the cross wire. For this purpose a fine wire or thread should be stretched across the centre of the slit. If the level of the collimator admits of alteration, it should be adjusted until the image of this thread coincides with the centre of the cross. If the collimator is fixed, the position of the thread on the slit must be altered until the condition is nearly satisfied, great accuracy of this adjustment not being essential.

Another method of focussing telescope and collimator is sometimes found more convenient, as it does not involve removing the prism or other apparatus from the table. This method is based on the fact that if a beam of light diverging from a point S falls on a prism, the distance of the virtual image S' from the prism is greater the greater the angle of incidence of the original beam. If the prism is placed in the position of minimum deviation (Fig. 86), the virtual image will be at the same distance as the object, but as the prism is turned in the direction of the hands of a watch, the virtual image will move further away. If the distance of S from the prism is very great, so that a parallel beam of light falls on the prism, the rays will remain parallel after refraction, whatever the angle of incidence. *If the original beam is parallel therefore, and only in that case, will the position of the prism have no effect on the distance of the virtual image from the observer.* This furnishes a delicate test of the parallelism of a beam of light. If the light is not parallel the adjustment may be made as follows*:

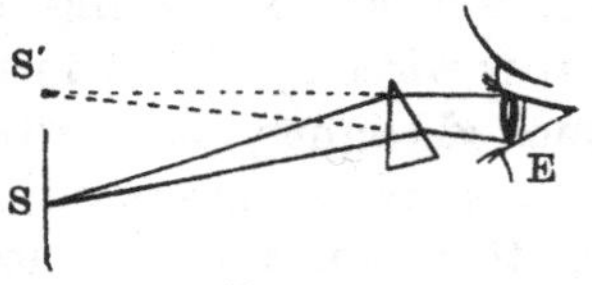

FIG. 86

Place a prism on the spectrometer table approximately in the position of minimum deviation. Illuminate the slit with sodium light and turn the telescope so as to see the sodium line. Fix the position of the telescope so that when the prism is turned round, the image moves across the field of view and comes to minimum deviation just at its edge or a little

* Schuster, *Phil. Mag.* VII, p. 95 (1879).

beyond. It is clear that there will now be two positions of the prism for which the sodium line will be in the centre of the field of view. In the first position, which we may call the "slanting position," the incidence of the beam on the prism is greater than at minimum deviation, while in the second, which we may call the "normal position," the incidence is smaller. Rotate the prism to the slanting position, and focus the telescope carefully until you see the edges of the slit quite sharply; it is better for this purpose not to have too narrow a slit. Now turn the prism into the normal position. If the focus is still good, the collimator supplies a parallel beam. But if the image is out of focus, this shews that the collimator is not in adjustment, and as we know that the change of position of the prism must have brought the image *nearer*, we may remove it to its old position by adjusting the *collimator*. When this is done the prism is turned back to the slanting position. In doing so we remove the image still further, and must focus again, but this time with the *telescope*.

Repeat the process several times, taking care always to focus the telescope when the prism is in the slanting position, and the collimator when the prism is in the normal position. After three or four changes it will generally be found that the image remains in focus during the displacement of the prism, and the adjustment is then complete.

If the telescope is now directed straight towards the collimator, the image of the slit should be in focus at the cross wires, and there should be no parallax on moving the eye to the right or left. When the prism is in use, if the faces of the prism are not quite plane, it may happen that the focus is slightly different according as light passes through the prism, or is reflected from one of its faces. Unless the image is rendered markedly indistinct it is best, however, not to alter the focus. The prisms generally supplied with spectrometers of ordinary size are sufficiently plane not to require any readjustment of focus.

We have seen that the line of sight of the telescope should be at right angles to the axis of rotation, *i.e.*, parallel to the plane of the divided circle.

This adjustment is more difficult to carry out. Fortunately the errors introduced when the condition is not accurately fulfilled are not very great, and unless great accuracy is required it need not be carried out. The method is instructive however, and should be understood.

In Fig. 87 let AB be the axis of rotation, LM the line of sight of the telescope. Also let HK be a piece of plane parallel glass, placed so that a ray of light passing along ML is reflected back along the same line LM, *i.e.* with HK at right

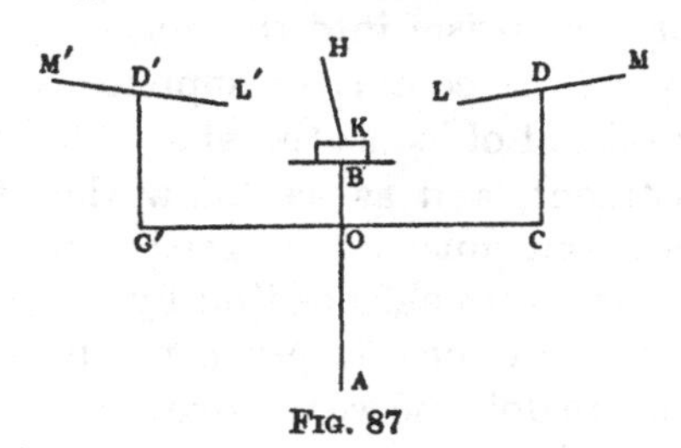

Fig. 87

angles to LM. If now the arm OCD is rotated through two right angles round AB, until the line of sight comes into the position $L'M'$, an inspection of the figure will shew that the surface HK will no longer be at right angles to $L'M'$, *unless the line LM was originally at right angles to AB.*

It remains to be shewn how the line of sight of the telescope may be adjusted to satisfy this condition. The telescope is provided with an eyepiece having an aperture at the side, through which light may fall on an inclined piece of glass inserted between the two lenses of the eyepiece. The light from a luminous burner is reflected from the glass plate, and illuminates the cross wires (Fig. 88).

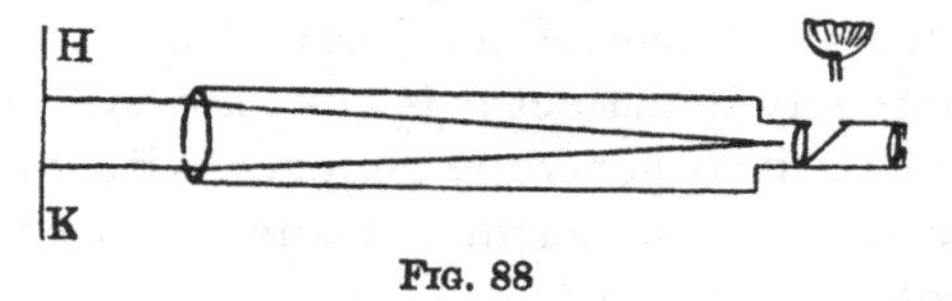

Fig. 88

The light diverging from these wires is converted by the object-glass of the telescope into a parallel beam of light, which is reflected from the plane surface HK along its own path, if that surface is at right angles to the line of sight of the telescope. An image of the cross wire will be formed which will coincide with the wires themselves if all the adjustments are perfect. A slight displacement of HK will displace the image of the cross wires, and unless HK is approximately in the right position, the image may not appear in the field of view at all. In making the adjustment proceed as follows. Place the glass plate on the table of the spectrometer so that the plane of the plate is perpendicular to the line joining two levelling screws of the table, and loosen the clamping screws of table and telescope. Turn the table until the line of sight of the telescope is, as far as you can judge by the eye, at right angles to the glass plate. Place your eye near the eyepiece of the telescope and look at the image of the telescope in the glass plate. This image should be in line with the telescope itself as in Fig. 89.

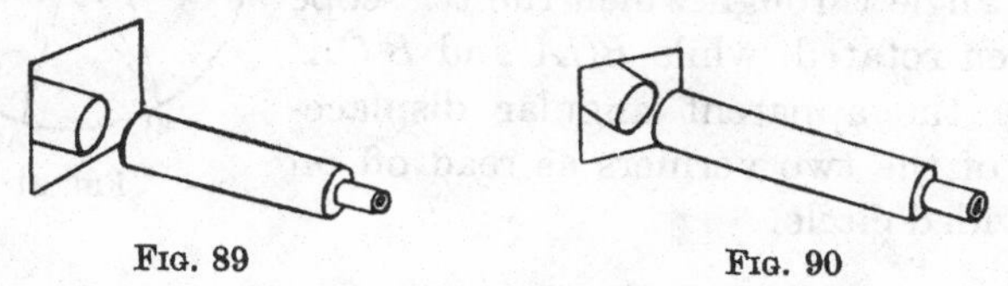

FIG. 89 FIG. 90

If the image points downwards, as in Fig. 90, the plate should be tilted backwards by means of the levelling screws. A little practice will enable you to set the plate nearly right in this way by eye. When this is done, clamp the table, place a light at the side of the eyepiece, and look through the telescope. A small angular movement of the telescope to the right or left should now bring the image of the cross wires into sight. Now level the glass plate till on moving the telescope the image of the cross wires coincides exactly with the wires themselves.

The adjustment having been made on one side, turn the telescope through two right angles, and see whether the image of the cross wires formed by reflection from the other side of the plate coincides with the wires. If this is the case, the line of sight is at right angles to the axis of rotation as required. If not, bring the two images into coincidence by first tilting the telescope until their distance apart has been halved, and then tilting the glass plate until the images are in coincidence. Turn the telescope again through two right angles, and if necessary repeat the adjustment till the images coincide with the cross wires in both positions of the telescope.

If the two images of the cross are not both sharply defined, the telescope has not been properly adjusted for parallel rays, or the glass plate is not perfectly plane.

Note.

In Fig. 91 let C be the centre of the divided circle. In any position of the telescope let one of the verniers be at A and the other at A', and let O be the axis of rotation of the telescope. Let the telescope be rotated till the verniers are at BB'. Then BOA will be the angle through which the telescope has been rotated, while BCA and $B'CA'$ will be the apparent angular displacements of the two verniers as read off on the divided circle.

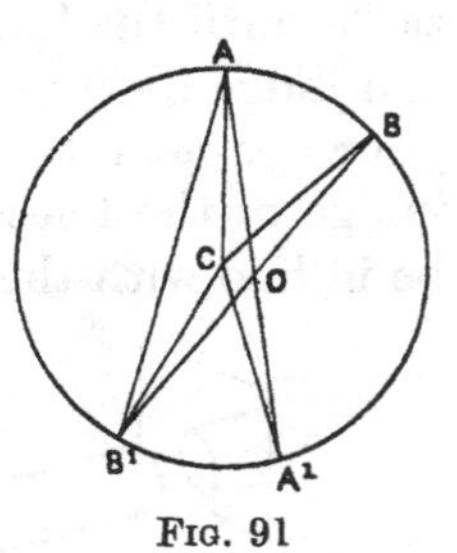

Fig. 91

But

$$BCA = 2OB'A = BOA + OB'A - OAB'$$

and

$$B'CA' = 2OAB' = BOA + OAB' - OB'A$$

Hence by addition $\overline{BCA + B'CA' = 2BOA}$

which shews that the angle through which the telescope is rotated is correctly obtained by taking the arithmetic mean of the angles read off on the vernier scale.

SECTION XXXIX

REFRACTIVE INDEX AND DISPERSION OF A SOLID BY THE SPECTROMETER

Apparatus required: *Spectrometer in adjustment, glass prism, small mirror and reading lens, sodium bead and flame.*

The method used in determining the refractive index of a solid is identical with that explained in the elementary exercise* on the same subject, with which the student is supposed to be familiar. The first step is to determine the refracting angle of the prism. For this purpose the prism must be placed on the table of the spectrometer so that its two refracting faces, and therefore their intersection, are perpendicular to the plane of the graduated circle. The table of the spectrometer ought to consist of a platform which can be levelled independently of the circular scale. If this is not the case, the prism should be placed on a separate support provided with levelling screws.

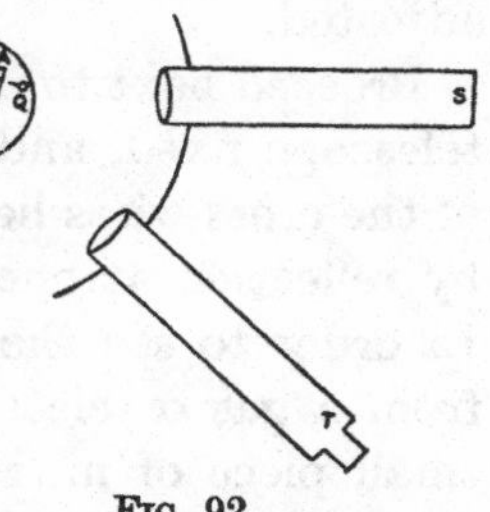

FIG. 92

In Fig. 92 PQR are the three levelling screws, and the prism is placed in the centre of the table in such a way that one of the faces, say AC, is at right angles to the line joining two screws, say R and Q. This may be done with sufficient accuracy by eye. To adjust the faces of the prism, place the collimator and telescope at a small angle to each other, illuminate the slit and turn the table of the spectrometer until an image of the slit, formed by reflection at AC, appears in the centre of the field.

The slit should be sufficiently narrow to allow you to place the cross accurately over its image; but it is not necessary to have it as narrow as you would use it in spectroscopic work,

* *Intermediate Practical Physics,* Sec. XXVIII.

unless you desire to resolve the double sodium line and measure the refractive index for each line separately.

The spectrometer table must now be levelled so that the image of a thread stretched across the centre of the slit coincides with the centre of the cross wires of the telescope. The table is then turned till the reflection of the slit is obtained from the face AB, and the image of the thread made to coincide with the centre of the cross wires by levelling the prism *by means of the screw* P, which is the only one that does not alter the inclination of the face AC, but only turns that face in its own plane. After the face AB has been levelled in this way, return to the face AC; if you have not originally succeeded in placing AC at right angles to QR, you may have slightly altered the inclination of the face. If so, set it right again by the screw Q which least disturbs the face AB.

Going backwards and forwards once or twice will always enable you to secure that the two faces are both properly adjusted.

Proceed next to measure the angle of the prism. Keep the telescope fixed, and turn the prism so that the intersection of the cross wires lies exactly on the image of the slit formed by reflection at one face of the prism. Read both verniers. In order to see the vernier divisions distinctly reflect light from a gas or electric lamp on to the scale by means of a small piece of mirror glass held near the scale. Next turn the prism till the image of the slip formed by the second face is on the cross wire, and again read both verniers. If the table has been turned through an angle $\theta°$ between the two observations, the angle of the prism is $180° - \theta°$.

If in moving from one position to the other a vernier passes over the zero of the scale, 360° must be added to the smaller reading, and the higher reading subtracted from it.

Alter the position of the telescope by a few degrees, and *repeat your observations*, obtaining a *second value* for the angle of the prism, which should agree closely with the first.

This method implies that the table of the spectrometer is moveable. If it is not so, the method described in the Note, p. 206, must be adopted.

Enter your observations as follows:

Date:

Spectrometer *A*, Prism *A*.

Determination of Angle of Prism.

	Vernier A.	Vernier B.
First position of prism ...	350° 6′ 0″	170° 2′ 20″
Second ,, ,, ...	110° 13′ 0″	290° 11′ 0″
Difference...	120° 7′ 0″	120° 8′ 40″
Mean	120° 7′ 50″	
∴ Angle of prism	59° 52′ 10″	

Similarly for the second measurement.

Having determined the angle of the prism proceed to find the minimum deviation for light of the kind for which the refractive index of the prism is required, *e.g.*, sodium light. Place a Bunsen burner with a sodium bead in it behind the slit, and without altering the adjustment of the prism turn the spectrometer table so that the light coming from the collimator is refracted through the prism. Find the refracted image of the slit in the first place by eye, and follow the image while the prism is slowly turned round till the position of minimum deviation is found roughly. When this is done bring the telescope into this position, and find the image of the slit in the telescope. Watch this image while the table of the spectrometer is slowly turned. If the direction of rotation is properly chosen, you will find the image moves slowly in the plane of deviation, comes to a standstill, and moves back again. Leave the prism in the position for which the deviation is least, place the cross wires of the telescope approximately on the image, clamp* the telescope and adjust the cross wires more accurately by means of the slow motion. Now turn the prism again backwards and forwards, making sure that the *centre* of the image of the slit just comes up to the centre of the cross wires, but does not pass beyond it. Adjust the telescope if necessary until you are quite satisfied that this is the case, *then clamp the table of the spectrometer*.

* See remarks as to clamping, &c., p. 192.

Now read both verniers. Remove the prism, unclamp the telescope, turn it so as to point directly towards the collimator, and adjust it until the image of the slit is bisected by the cross wires. Read the verniers to determine the position of the telescope when there is no deviation of the ray. The difference in the readings in this position and in the position of minimum deviation gives the angle of minimum deviation.

Obtain two independent readings of the position of minimum deviation so as to secure accurate results.

When the spectrometer table is fixed, it is not material whether the direct reading is taken before that of minimum deviation or *vice versâ*. *But when the graduated circle moves with the spectrometer table the above order must be adhered to.* For it has already been pointed out that correct values will in that case only be obtained if either the telescope only, or the table only, is turned round between two readings. As the adjustment to minimum deviation necessarily involves the turning round of the table, this should be done first and the table be *clamped* before the telescope is moved round to take the direct reading.

The observations are entered as follows:

Determination of Minimum Deviation.

	Vernier A.	Vernier B.
Prism at minimum deviation	143° 40′ 00″	323° 40′ 40″
Direct reading	94° 13′ 00″	274° 14′ 20″
	49° 27′ 00″	49° 26′ 20″

Angle of minimum deviation... 49° 26′ 40″

Similarly for the second experiment.

Angle of minimum deviation... 49° 27′ 40″

Mean D		49° 27′ 10″
Angle of prism α ...		59° 52′ 10″
$D + \alpha$	=	109° 19′ 20″
$(D + \alpha)/2$	=	54° 39′ 40″
$\alpha/2$	=	29° 56′ 05″

$$\mu_D = \frac{\sin \dfrac{D+\alpha}{2}}{\sin \dfrac{\alpha}{2}} = 1{\cdot}6347.$$

The observations should be sufficiently accurate to give the third decimal place with certainty, and the fourth decimal place with fair accuracy. The numerical calculation should be carried out with a table of seven figure logarithms.

Calculate the error produced in n on the supposition that an error of half a minute has been made (*a*) in the measurement of the angle, (*b*) in the determination of the minimum deviation.

Repeat the determination, using lithium and strontium flames, and draw a curve connecting the indices of refraction (ordinates) with the wave numbers (abscissae).

Taking $\lambda_0 = 2700$ draw a second curve connecting the indices with Hartmann's numbers $1/(\lambda - \lambda_0)$.

If this does not give a straight line find some more suitable value for λ_0 and plot a new curve.

From the last curve determine $\mu_F - \mu_C$, where F and C are the green and red H lines of the table, p. 191, and calculate the dispersion of the glass of the prism $(\mu_F - \mu_C)/(\mu_D - 1)$ and the reciprocal dispersion $(\mu_D - 1)/(\mu_F - \mu_C)$. The latter varies for different glasses from 20 to 70*.

Resolving Power.

Determine the resolving power of the instrument by measuring the difference δ of deviation it produces in the two sodium lines if they are just resolved by it, the slit being as fine as possible.

Place stops of various diameters from 2·0 to ·5 cm. in front of the objective of the telescope and measure the diameter d of the smallest stop which just allows the lines to be seen as distinct lines. The resolving power is $d\delta/5{\cdot}9 \times 10^{-8}$, where δ must be in circular measure and $5{\cdot}9 \times 10^{-8}$ is the difference of wave lengths of the two sodium lines.

Record as follows:

Spectrometer *A*. Prism *A*.

Mean deviation for 1st ray	...	49° 27′ 0″
,, ,, 2nd ,,	...	49° 27′ 20″
Difference	...	20″

* See *Dict. App. Physics*, vol. IV, "Optical Glass."

Difference circular measure $= \delta$		$= \cdot 000095$
Width of stop, d	...	$\cdot 70$ cm.
Resolving power observed...	...	1100
,, ,, theoretical	...	1000
Free diameter of objective	...	$2 \cdot 2$ cms.

$$\therefore \text{ Resolving power of instrument } = \frac{2\cdot 2}{\cdot 7} 1100$$

$$= 3400$$

Note.

If the table of the spectrometer is not moveable, the angle of the prism may be measured as follows: Place the prism with its refracting edge towards the collimator in such a position that about half the beam of light falls on one and half on the other face of the prism. A reflected image may now be obtained from each face, and if the telescope is first placed so as to point to one image, and then turned until it points to the other, the angle through which the telescope has been turned is twice the angle of the prism. In Fig. 93, let I be the virtual image of the slit formed by the collimator, ABC the prism, I_1 and I_2 the two images formed by reflection at the faces AB and AC of the prism. By the laws of reflection we have the angles

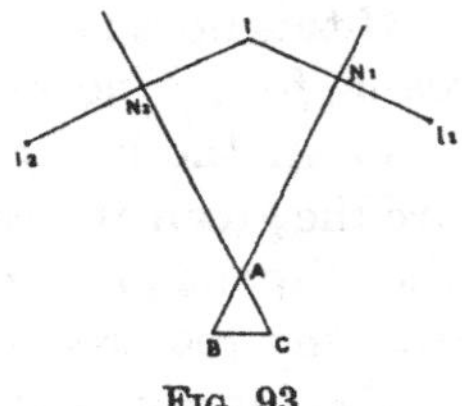

FIG. 93

$$IAI_1 = 2IAN_1$$

$$IAI_2 = 2IAN_2$$

$$\therefore I_1AI_2 = 2N_1AN_2 = 2\alpha.$$

The images I_1 and I_2 therefore subtend an angle 2α at the refracting edge A, and if the telescope is pointed first towards I_1 and then towards I_2, the angle through which it turns will be 2α, if the axis of rotation of the telescope is at A, or if I is at an infinite distance. Hence, in order that this method should give correct results, the collimator ought to be well adjusted for parallel light, and to correct any error due to a faulty adjustment the refracting edge should be placed nearly over the axis of rotation of the telescope. The errors introduced may otherwise be quite appreciable. If, for instance, the distance between the refracting edge and the axis of rotation is one centimetre and the image of the slit at a distance of 50 metres, which is quite possible if the collimator is of ordinary size, the error may amount to over a minute of arc.

Quite apart from this source of error, which might be avoided, the method is not a good one, for each image is formed by an unsymmetrical beam of light, filling at the utmost only half the telescope lens, and serious errors may creep in due to the aberrations of the lenses.

SECTION XL

REFRACTIVE INDEX AND DISPERSION OF A LIQUID. SPECIFIC REFRACTIVE POWERS

Apparatus required: *Spectrometer, hollow glass prism, alcohol, thermometer, Bunsen flame, and sodium bead.*

When the index of refraction of a liquid is to be determined, the liquid is placed in a hollow prism the vertical sides of which are two plates of glass. If these glass plates are accurately plane parallel they produce no deviation of the rays of light passing through them, and the deviation observed is due to the liquid only. Hence the angle of the prism of liquid may be measured, and the refractive index determined by the deviation, just as in the case of a solid glass prism. If the plates of glass are slightly prismatic their effects on the apparent angle and deviation must be eliminated. To enable this to be done they are not cemented to the rest of the prism, but are made moveable, and are kept in position by rubber bands or by clips. After observations of the angle of the prism and of the deviation have been made each plate is rotated about its normal through 180° and the observations repeated. The means of the two sets of observations of the angle and of the deviation are the required angle and deviation of the liquid prism.

Clean the prism provided, and fill it with absolute alcohol at the temperature of the room.

As the prism may leak a little, it is not placed directly on the table of the spectrometer, but in a shallow metal dish which stands on three short legs on the table, and itself supports the prism at three points.

The adjustment of the prism and the determination of the angle and deviation are made as in the previous section. In determining the angle the stronger of the two reflected images of the slit seen in the telescope should be used in each case. It is formed by reflection at the air-glass surface, while the weaker is formed by reflection at the glass-liquid surface.

The glass plates forming the sides of the prism are then rotated as described above and the observations repeated. Each plate is provided with a mark, the position of which should be recorded, along with the observations.

The mean deviation D and the mean angle α are then found and the refractive index calculated from them in the usual way.

Enter your observations and results as follows:

Date:

PRISM II. ABSOLUTE ALCOHOL.

Sodium light.

Determination of deviation produced by prism when the marks on the glass plates are near angle of prism:

	Vernier A.	Vernier B.
With prism	229° 5′ 0″	49° 4′ 30″
Without prism ...	23° 9′ 0″	203° 7′ 30″
	25° 56′ 0″	25° 57′ 0″
Apparent deviation	25° 56′ 30″	

Determination of angle of prism when the marks on the plates are near angle of prism:

First position ...	246° 29′ 30″	66° 29′ 30″
Second ,, ...	126° 29′ 0″	306° 29′ 30″
Difference	120° 0′ 30″	120° 0′ 0″
Mean...	120° 0′ 15″	
∴ Apparent angle of prism =	59° 59′ 45″	

Give similarly the readings of the verniers when the plates have been rotated and the marks are near the base of the prism, and collect the results as follows:

Mean minimum deviation = D	25° 36′ 30″
Mean angle of prism = α	59° 51′ 42″
∴ Refractive index μ_D	1·3635
Temperature of alcohol	18° C.
Refractive index calculated from refractive powers	1·359

If greater accuracy is required several measurements should be made and the mean of the results taken.

Substitute lithium and strontium for sodium and determine the indices for the red and green lines.

Draw curves as in the case of the glass prism and determine from them the reciprocal dispersion of ethyl alcohol. Its value for liquids generally lies between 10 and 60.

The observed refractive index may be compared with the index calculated by a method founded on the following facts:

If μ is the refractive index and d the density of a substance, the value of the expression $\frac{\mu - 1}{d}$ is called by Gladstone the specific refractive power of the substance. It is not sensibly altered by variations of temperature or pressure, but changes when the substance passes from the liquid to the gaseous state. Lorenz and Lorentz have from theory deduced another expression, namely, $\frac{\mu^2 - 1}{\mu^2 + 2} \cdot \frac{1}{d}$, which remains nearly constant, even when the substance changes its state. For the purpose of this exercise it will be sufficient to use the former expression, as it more easily lends itself to numerical calculation. The refractive power varies with the wave length of the light used, and we shall take μ to refer to sodium light.

It is found by experiment that the refractive power of a mixture of substances is very nearly equal to the mean refractive power of the constituents. Thus, if p_1, p_2, p_3, &c. grams. of a number of liquids are mixed together, we have very nearly:

$$P\frac{N - 1}{D} = p_1 \frac{\mu_1 - 1}{d_1} + p_2 \frac{\mu_2 - 1}{d_2} + p_3 \frac{\mu_3 - 1}{d_3} + \&c.,$$

where $P = p_1 + p_2 + p_3 + \ldots$, and $(N - 1)/D$ is the refractive power of the mixture.

It has also been found, that for a compound, the molecule M of which contains n_1 atoms of an element the atomic weight of which is m_1, n_2 atoms of an element the atomic weight of which is m_2, and so on, the weights of the different elements in the combination being therefore n_1m_1, n_2m_2, &c., we have very nearly:

$$M.\frac{\mu - 1}{d} = n_1m_1\frac{\mu_1 - 1}{d_1} + n_2m_2\frac{\mu_2 - 1}{d_2} + \&c.$$

If we call $m_1\frac{\mu_1 - 1}{d_1}$ the atomic refractive power of the first element, and designate it by r_1, and so on for the other elements, and $M\frac{\mu - 1}{d}$ the molecular refractive power R of the compound, we have then the equation

$$R = n_1r_1 + n_2r_2 + \&c.,$$

or in words: the refractive power of a compound is the sum of the products formed by multiplying the atomic refractive power of each element by the number of atoms of that element contained in the molecule of the compound.

We cannot directly determine the atomic refractive power of an element entering into a compound, but we may find it indirectly. If, for instance, we take three compounds containing carbon, hydrogen, and oxygen in different proportions, we obtain three equations to determine r_1, r_2 and r_3, the atomic refractions of carbon, hydrogen, and oxygen respectively, and if we find that the refractive indices of other compounds may be determined with the help of the values so found, we shall have justified the above statement. It appears, however, that the way in which an atom is combined affects its refractive power, and to obtain consistent results it is necessary to assume, for instance, that oxygen in the carbonyl group has a different refractive power from the hydroxyl oxygen.

Calculate the refractive index of ethyl alcohol C_2H_6O, making use of the following numbers.

Atomic refraction of H (H = 1)	...	1·47
,, ,, of O in alcohols (O = 16)		2·65
,, ,, of C (C = 12)	...	4·71
Density of ethyl alcohol, at 20° =	...	·788
Coefficient of cubical expansion =	...	·0010
Calculated index of alcohol at 18° C. =		1·359
Measured ,, ,, =		1·363

SECTION XLI

PHOTOMETRY

Apparatus required: *Photometer, scale, sperm candles or a standard electric lamp, incandescent lamp, voltmeter, ammeter and adjustable resistances.*

The photometer is an instrument for comparing the illuminating powers of different sources of light. One of the most accurate forms of the instrument is that in which the two sources are placed on opposite sides of a screen capable of partly reflecting and partly transmitting the light from each source. When this screen is placed in such a position that the straight lines drawn from it to the two sources make equal angles with the screen, and the illumination of the two sides of the screen appears the same, the amounts of light to be compared are to each other as the squares of the distances of the sources from the screen.

The standard source is the flame of a sperm candle, $\frac{7}{8}$ inch diameter, consuming 120 grains of sperm per hour, and the standard direction any line in a horizontal plane. If the consumption of sperm differs from the standard rate by less than 10 grains per hour, the illuminating power may be taken as proportional to the consumption.

Candles are too variable to provide an accurate standard and either an incandescent electric lamp, with its filament in one plane, standardised and run at a stated voltage, or a 10-candle Harcourt pentane lamp, is generally substituted. In what follows it is supposed that an incandescent lamp which has been compared with a standard is available.

Arrange the secondary standard, the photometer, and the second incandescent lamp provided, in the order named, on the photometer bench which consists of two graduated rods

3 metres long on which the supports carrying the incandescent lamps and photometer slide or run. The standard may be fixed with its filament over the zero of the scale and with the plane of its filament perpendicular to the rods, and the photometer at 100 cms. The second lamp is placed on the other side of the photometer with the plane of symmetry of its filament perpendicular to the base and its centre over the fixed mark on its stand (Fig. 94). Connect each lamp through

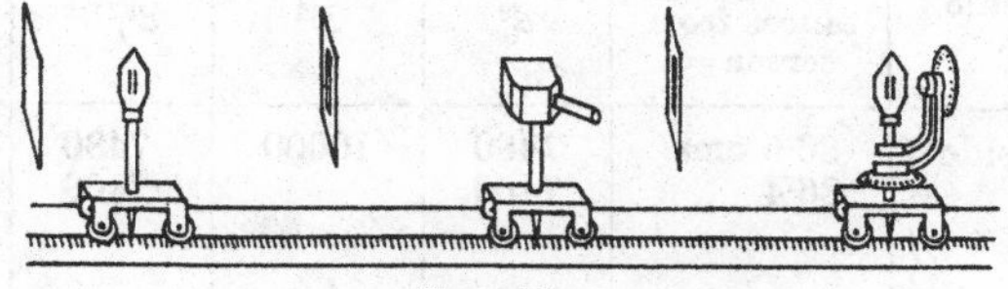

Fig. 94

a resistance to the light mains or a storage battery and adjust the resistance till a voltmeter across the lamp terminals shews that the proper electromotive force for which the lamp is marked is applied to it. Check this from time to time as the observations proceed. The second lamp can be moved backwards and forwards along the scale till the illumination of the two sides of the photometer screen is equal. If the simple Bunsen screen is used, this is the case when the semi-opaque and semi-transparent parts of the screen appear of equal brightness, and if the Lummer-Brodhun instrument* is used, when the parts of the field of view of the telescope through which the screen is viewed appear of equal brightness.

If the photometer screen can be reversed this should be done and equality of illumination again secured by moving the second lamp. At each equality the position should be read on the scale. These observations should be repeated three or four times and the means taken.

Rotate the second lamp so that its plane of symmetry is now inclined at 45° to the line drawn to the photometer, and repeat the observations. Rotate the lamp again through 45° and repeat. Continue the rotations and observations till those first taken have been repeated.

* This is the instrument shewn in Fig. 94.

Record as follows:

Date:

Secondary standard *B*, 200 volts, 40 candles.

Secondary standard at ... ... 0 cms.
Photometer at ... ... 100 „
$\therefore D =$ ... ... 100 „

80 candle lamp	Mean distances from screen $= d$	d^2	D^2	d^2/D^2	Candle power
Broadside	86·5 cms.	7480	10000	·7480	29·9
	86·4 „	7460		·7460	
45°					
90			&c.		
135					
180					

Represent these results graphically as follows:

Draw a plan of the filament as seen from above, with its plane of symmetry and lines inclined at 45° and 90° respectively to it. Along each line take a length proportional to the candle power observed in its direction. Join the points so determined by a smooth curve. This is known as a "horizontal photometer curve" for the flame. By taking observations in directions inclined at intermediate angles it can be drawn with greater accuracy.

With the lamp in the broadside-on position rotate it about a horizontal axis through the centre of the filament and perpendicular to the rods. Take observations of candle powers in directions inclined at 45° and 90° to the plane of symmetry and record as above.

With the lamp broadside-on again determine the candle power, reading the current taken by the lamp.

Reduce the electromotive force in the lamp by 5 %, read the volts and amperes and determine its candle power. Repeat with reductions of 5 % at each repetition.

Draw a curve with candle powers as ordinates and electromotive forces applied as abscissae.

Calculate the watts used in each case and plot a second curve shewing how the candle power varies with the watts used.

Record as follows:

Volts	Amperes	Watts	d	d^2	D^2	d^2/D^2	Candle power
240	·125	30·1	86·5	7480	10000	·748	29·9
230	·121	27·8	81·2	6600		·660	26·4
220							19·8
...	...	&c.	...	...	...	...	...
120	·067	8·0	21·9	480		·048	1·2

SECTION XLII

INTERFERENCE OF LIGHT

Apparatus required: *Bunsen flame, sodium bead, slit, biprism, micrometer microscope, metre scale.*

When two beams of homogeneous light coming from the same source cross each other after having described paths differing in length, the vibrations due to the two may be in opposite directions and neutralise each other at certain points of the region where the beams cross. At such points the joint action of the two beams will produce partial or total darkness, and if a screen is placed in this region, a series of light and dark "interference" bands will be seen on it.

In the case in which the source of light is a narrow vertical slit behind which a sodium flame is placed, and the two beams are produced by the passage of the light from this slit through a double prism the section of which is indicated in Fig. 95, the distance x between consecutive dark bands on the screen, is related to the wave length λ of the light, the distance a of the slit from the screen, and the distance apart c of the two virtual images of the slit formed by the biprism, by the equation: $\lambda = \frac{xc}{a}$*.

FIG. 95

This method of producing interference bands may therefore be used to determine approximately the wave length of light.

The experiment is performed on an "optical bench," a simple form of which is shewn in Fig. 96.

The slit consists of two brass plates with straight bevelled edges, attached by screws to a wooden stand provided with levelling screws.

* Schuster, *Optics*, chap IV. Fresnel's experiments. The deviation of each ray is $(\mu_1 - \mu_0)\,a$, where a is the refracting angle of the prism through which it passes, μ_1 the index of the prism, μ_0 of the surrounding medium.

The brass plates should be adjusted so that their edges are about ·1 mm. apart and parallel to each other. This is most easily done by placing a strip of thin writing paper between the edges, pressing the plates together and screwing them firmly to the wood, and then removing the paper.

The biprism is mounted on a similar stand.

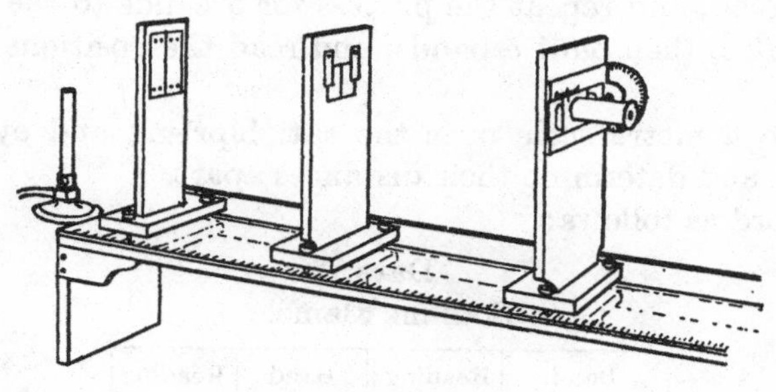

FIG. 96

The micrometer microscope consists of an eyepiece magnifying a few times, across the focal plane of which a cross wire, or by preference a pointed vertical needle, can be moved by means of a screw with a graduated head on which the number of turns can be read. The arrangement is mounted on a heavy wooden base.

The three stands described above slide along a wooden bench provided with a groove in which two levelling screws of each stand slide, and with a graduated scale.

Place the slit vertical at one end of the stand, and the sodium flame behind it, about 15 cms. in front place the biprism, and about 45 cms. in front of the biprism the micrometer eyepiece. If the slit, edge of biprism, and centre of field of view of the eyepiece, are in a line, and the edge of the prism is nearly parallel to the slit, a series of light and dark vertical bands should be seen on looking through the eyepiece. If no bands are visible move the biprism backwards and forwards along the bench and tilt it slightly in its own plane till bands are visible. Now watch the bands and continue the tilting by means of the single screw at one side of the biprism stand. Stop the tilting when the bands are most distinct. The edge of the biprism will now be parallel to the slit. The biprism and eyepiece should now be moved along the bench till the bands are both distinct and at a convenient distance apart for measurement. When this is the case adjust the reference mark in the focal plane of the eyepiece to the centre of a light band, read the screw head, move to the next band, read, and repeat the process for 5 bands to the left of the centre, then omit 5 bands, and read the positions of the next 5.

Place a metre scale over the slit, biprism, and eyepiece stands, and determine their distances apart.

Record as follows:

Date:
Sodium Flame.

Band	Reading	Band	Reading
1	37·72	11	41·48
2	38·12	12	41·85
3	38·54	13	42·20
4	38·84	14	42·59
5	39·25	15	42·93
Mean	38·494		42·210
Difference		3·716	

Difference for 10 bands ... ... = 3·716 scale divisions.
1 scale division ... ... = ·05 cm.
∴ Distance apart of bands ... = ·0186 cm.

To determine the distance apart of the two virtual images of the slit formed by the biprism, place a lens of about 12 cms. focal length, provided with a stop of about ·5 cm. diameter, in such a position between the biprism and eyepiece that the two images of the slit are in focus. Measure the distance of the images apart by the micrometer, and the distances of the lens from slit and eyepiece by the metre scale. If a second position of the lens can be found for which the images are again in focus repeat the observations.

The virtual images since they are formed by prisms of small refracting angles for nearly normal incidence are at the same distance from the prism as the object. If therefore u is the distance of the slit from the lens, it will also represent the distance of the virtual images from the lens. The measured distance between the real images in the focal plane of the eyepiece being c_1, the distance between the virtual images will be given by $c = cu_1/v$, when v is the distance between the lens and the focal plane of the eyepiece. For a second position of the lens, if c_2 is the observed distance between the images, $c = c_2 v/u$.

Either of these observations will give c, but the measurement of u and v may be avoided altogether, for if we multiply the equations together we have $c = \sqrt{c_1 c_2}$.

Arrange observations and work as follows:

Distance of lens from		Distance of images apart		
slit $=u$	screen $=v$	Scale divisions	cms.	c
19·7 cm.	40·3 cm.	7·80	·390	·191
40·2	19·8	1·93	·0915	·189
			$\sqrt{c_1 c_2}$	·189
			Mean	·190

$u + v = 60{\cdot}0$ cms. $= a$.

$$\therefore \lambda_D = \frac{xc}{a} = \frac{{\cdot}0186 \times {\cdot}190}{60} = {\cdot}000{,}0589 = 5{\cdot}89 \times 10^{-5} \text{ cms.}$$

Known wave length of sodium light $= 5{\cdot}893 \times 10^{-5}$ cms.

Place a small piece of thin plane parallel glass in contact with the obtuse angle of the prism. Notice that it has no effect on the interference fringes. Put a drop of water between the glass and prism and notice that the distance between the fringes has increased. Measure their distance apart and calculate the index of refraction of the glass of the biprism, given that the distances apart of the fringes are inversely proportional to $\mu_2 - \mu_1$, where μ_2 is the index of the glass of the biprism, μ_1 that of the medium between the biprism and plane plate.

Record as follows:

Biprism *B*.

Mean distance apart of fringes with air	...	36·10
" " " water	...	100·8

$$\frac{\mu_2 - 1}{\mu_2 - 1{\cdot}333} = \frac{100{\cdot}8}{36{\cdot}1}.$$

$$\therefore \mu_2 = 1{\cdot}519.$$

Once μ_2 is known the index of any other liquid substituted for the water may be found.

SECTION XLIII

NEWTON'S RINGS

Apparatus required: *Newton's Rings apparatus, sodium flame, measuring microscope.*

The apparatus provided consists of a lens A of about 15 cms. focal length by means of which a beam of light diverging from a sodium flame is rendered parallel. The parallel beam is reflected by a plane glass plate B on to a glass plate D pressed against the upper surface of a lens C. On reflection the light passes through the glass plate B to the sliding microscope E, by means of which the interference rings produced by light reflected at the under surface of the plate D and the upper surface of the lens C respectively are measured.

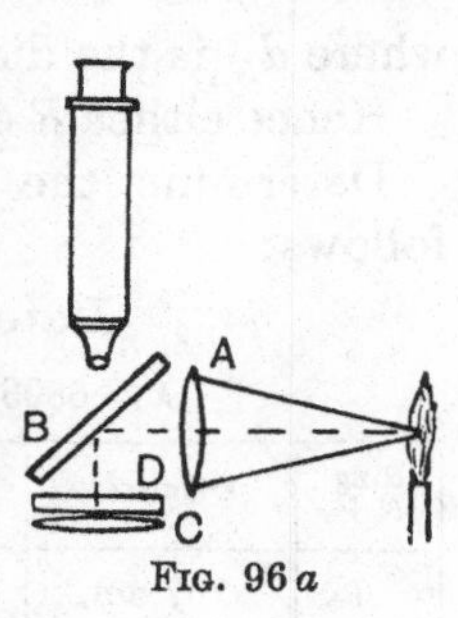

FIG. 96 *a*

By means of the screws in the plate which secures the lens C and plate D bring the system of rings seen by reflection into the middle of the lens and place the microscope so that the rings are in the middle of the field.

Measure the diameters of the first 10 dark rings, and tabulate as shewn below.

The condition that the rays reflected at a point on the under surface of the plate D distant r from the point of contact may reinforce the rays reflected from the top surface of the lens at a distance t below is

$$2.t + \frac{\lambda}{2} = 2N\frac{\lambda}{2},$$

where N is an integer.

But $t\,(2R - t) = r_N{}^2$, or, since t is small,

$$t = \frac{r_N{}^2}{2R}, \text{ and } \frac{r_N{}^2}{R} = (2N - 1)\frac{\lambda}{2}.$$

As it is not possible to ensure that plate and lens are optically in contact we do not know the value of N for any ring, but if we increase N by n where n is a convenient known number we again have

$$r^2_{N+n} = (2(N+n) - 1)\frac{\lambda}{2}.$$

Hence
$$\frac{r^2_{N+n} - r_N^2}{R} = 2n\frac{\lambda}{2},$$

or
$$R\lambda = \frac{d^2_{N+n} - d_N^2}{4n},$$

where d_N is the diameter of any ring.

Hence either R or λ can be calculated if the other is known.

Determine the Radius of Curvature R. Recording as follows:

Date:

$\lambda = 5893 \times 10^{-8}$ cms. $n = 5$. $4n\lambda = \cdot 0001178$.

Ring $N+$	Diameter	(Diameter)^2	Ring $N+$	Diameter	(Diameter)^2	Diffn. of (diam.)^2
1	·167 cm.	·0282	6	·247 cm.	·0610	·0330
2	·185	·0342	7	·259	·0670	·0328
3	·204	·0410	8	·271	·0734	·0324
4	·220	·0484	9	·283	·0800	·0316
5	·234	·0547	10	·294	·0864	·0317
					Mean	·0323

Hence
$$R = \frac{\cdot 0323}{\cdot 0001178} = 27^4 \text{ cms.}$$

SECTION XLIV

WAVE LENGTH OF LIGHT BY THE DIFFRACTION GRATING

Apparatus required: *Spectrometer, diffraction grating, Bunsen flame, and sodium bead.*

When a beam of homogeneous light diverging from a point passes through a transparent plate on one surface of which a series of equidistant fine parallel lines has been ruled, the emergent light appears to diverge from a number of virtual images, one of which is coincident with the luminous point, the others, which lie on either side of the luminous point, are called diffraction images. The surface on which the lines are ruled is called a diffraction surface or grating.

In the case of a thin parallel plate of a transparent medium with equidistant parallel lines ruled on one surface, if a parallel beam of light is incident normally on the ruled surface each diffracted beam will be a parallel one, and will be deviated by the angle θ, where $n\lambda = b \sin \theta$, n being an integer called the order of the image, and b the distance of the lines apart measured from centre to centre. Hence if the grating is placed on the table of a spectrometer, the collimator of which furnishes a beam of parallel light, and the telescope of which is focussed for parallel rays, the whole of the images will in turn be visible through the telescope as it is moved from one side to the other. The deviation of each diffracted beam may therefore be determinated on the graduated circular scale of the instrument. If a narrow illuminated slit is used instead of a luminous point, the images will be linear.

Adjust the spectrometer as described in Section XXXVIII. Place a sodium flame behind the slit. Using the Gauss eyepiece in the telescope direct it towards the collimator, adjust the centre of the cross wires to the image of the slit, and clamp the telescope.

Place the diffraction grating on the table of the instrument with the plane of the ruling perpendicular to the line joining two of the screws of the table.

Rotate the table till the image of the cross wires formed by reflection at the surface of the grating is seen in the field of view. Adjust the table by means of the levelling screws till the reflected and direct images of the wires coincide. The grating is then perpendicular to the axis of the collimator and telescope.

Remove the Gauss and substitute an ordinary eyepiece. Rotate the telescope till the first diffracted image is on the cross wires. If it is indistinct the screw of the table which will tilt the lines of the grating in their own plane should be rotated till the image is as distinct as possible. The width of the slit should then be varied till the two sodium lines are resolved quite distinctly.

Adjust the telescope by means of the tangent screws till the centre of the cross wires coincides with one of the lines, and read the vernier of the circular scale. Adjust to the other line and again read. Now rotate the telescope, adjust to the direct image and read the circular scale. Then rotate further in the same direction till the first diffracted image on the other side is seen, adjust and read the scale for each line. Again return to the direct image and take readings. By rotating still further in each direction, the 2nd, 3rd, &c. images may be found and their positions read.

To obtain the wave length of light the distance b between the lines must be known. In an absolute determination the most difficult part of the measurement consists in finding this distance. It may be determined by comparing it by means of a microscope, with a stage micrometer scale graduated in ·01 mm., using a high power objective, then comparing the micrometer scale with a millimetre division on a standard metre scale, using a low power objective. In the present exercise b will be assumed to be known.

Record the observations as follows:

Date:

Spectrometer A. Grating A.

Most refrangible Sodium line.

	Vernier A	Vernier B
First order, right	202° 11′ 30″	22° 11′ 00″
Direct reading	179 10 30	359 10 00
First order, left	156 9 30	336 9 30
Deviation, right	23 1 00	23 1 00
,, left	23 1 00	23 0 30
,, mean	23 1 00	23 0 45

Mean deviation for first order spectra ... 23° 0′ 52″.

Similarly for the other line, and for the spectra of the 2nd and higher orders.

Tabulate results as follows:

Grating space of A ... $= b = \cdot 00015062$ cm.

n	Angle of deviation $=\theta$	$\sin\theta$	$\frac{\sin\theta}{n}$	$b\frac{\sin\theta}{n} = \lambda$
1	23° 0′ 52″	·3909	·3909	·000,05888
2	51 25 0	·7817	·3908	·000,05887
			Mean	·000,05888

If the telescope is now turned to one of the first order images, and the table on which the grating is placed is slowly rotated, it will be seen that the position of the image, and therefore the deviation, change. A position of the grating can be found for which the deviation is a minimum. Adjust the telescope on the image when least deviated, and read the verniers. Keeping the table fixed, rotate the telescope till the direct image is seen. Read the verniers. Now rotate table and telescope till the minimum position of the first image on the other side is found, and repeat the observation of the deviated and of the direct image.

The minimum deviation ϕ is connected with λ and b by the equation $2b \sin\frac{\phi}{2} = n\lambda$, hence λ may again be determined.

Tabulate as follows:

First order spectra.

		Vernier A	Vernier B
Deviated reading right		202° 42′ 0″	22° 42′ 30″
Direct ,, ,,		180° 8′ 30″	0° 10′ 0″
Deviation to right ...	=	22° 31′ 30″	22° 32′ 30″
Mean	=	22° 32′ 0″	
Deviated reading left	=		
Direct ,, ,,	=	&c.	
Deviation to left ...	=		
Mean	=		

Mean deviation for first order spectra = 22° 32′ 15″.

Similarly for the second order spectra, collecting the results in tabular form as follows:

n	ϕ	$\frac{\phi}{2}$	$\sin\frac{\phi}{2}$	$\frac{1}{n}\sin\frac{\phi}{2}$	$\frac{2b}{n}\sin\frac{\phi}{2}=\lambda$
1	22° 32′ 15″	11° 16′ 7″	·1954	·1954	·000,05887
2	46 2 0	23 1 0	·3910	·1955	5889
				Mean	·000,05888

By using the grating so that the diffraction images are formed by rays reflected at the ruled surface, further determinations of λ may be made.

Resolving Power.

As in the case of the prism spectroscope the resolving power of a grating spectroscope is defined as the quotient of the wave length of the line of light used, by the least change of wave length which the instrument shews as a line distinct from the former.

It is shewn in text-books that this is equal for a grating spectroscope to the product of the order of the spectrum n into the total number of lines N of the grating crossed by the beam entering the telescope.

Determine the resolving power by placing stops of diameters from 2·0 cms. to ·2 cm. in front of the objective of the telescope till at diameter d the sodium lines $\lambda = 5890{\cdot}0$ and $5895{\cdot}9$ are only just resolved in the spectrum of the first order, *i.e.* for $n = 1$, with normal incidence. Then if θ is the deviation $N = \dfrac{d}{b\,.\cos\theta}$ and the resolving power $= \dfrac{nd}{b\cos\theta}$; and should theoretically be 1000. From the observed resolving power with the stop calculate the resolving power of the spectroscope when the beam from the grating fills the whole diameter D of the objective.

Record as follows:

Grating A, $b = 1{\cdot}5 \times 10^{-4}$, $\theta = 23°$.

First order spectrum.

Least diameter for resolution $d = {\cdot}25$ cm.

$$\text{Resolving power with stop} = \frac{{\cdot}25}{1{\cdot}5 \times 10^{-4} \times {\cdot}92} = 1800.$$

Free diameter of objective ... ... 2·2 cms.

∴ Resolving power of instrument in first order spectrum

$$= \frac{2{\cdot}2}{{\cdot}25} \times 1800 = 15{,}800.$$

SECTION XLV

ROTATION OF PLANE OF POLARISATION

Apparatus required: *Two Nicol's prisms, two tubes with glass ends, Rochelle salt, a lithium and a sodium flame.*

When a beam of homogeneous plane polarised light is transmitted through certain solids, liquids, and solutions, it is found that the plane of polarisation is rotated through an angle R proportional to the length l of the path of the ray in the substance, to the density d of the substance, and dependent on the nature of the substance and of the light used; *i.e.* $R = d.l.\rho$, where ρ is known as the specific rotatory power of the substance for the light used. In general ρ increases as the wave length of the light used decreases, hence the necessity for homogeneous light in making measurements.

The object of the present section is to verify the laws of rotation of the plane of polarisation of light by solutions of certain substances in a liquid which does not itself produce rotation. In this case the rotation is very nearly proportional to the mass of dissolved substance per c.c. of the solution. The rotation produced by a layer 1 cm. thick of a solution containing 1 gram per c.c. is called the specific rotatory power of the substance dissolved, for the kind of light used.

If a solution contains in 1 c.c. a grams of a substance the specific rotatory power of which is ρ, and a beam of homogeneous plane polarised light passes through l cms. of the solution, and R is the rotation of the plane of polarisation produced, then $R = a.l.\rho$.

Take about 75 c.c. of water, add to it 20 grams of Rochelle salt $KNaC_4H_4O_6$. When the salt has dissolved add water till the volume = 100 c.c.

If Rochelle salt is not available add 10 grams of sugar to 85 c.c. of water. When the sugar has dissolved dilute the solution till its volume = 100 c.c.

If the sugar solution is coloured, mix with it about 2 grams of bone black, allow it to stand a few minutes and then filter. If the filtrate is then bright and clear it may be used.

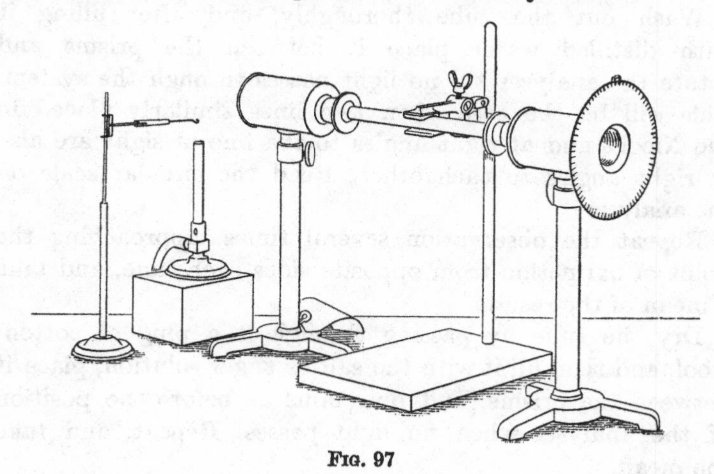

Fig. 97

Arrange the solution tube provided, which is a glass tube closed at the ends by moveable plane glass discs, and two mounted Nicol's prisms, in such a way that the light of a Bunsen flame containing a bead of a sodium salt can be seen through the tube and prisms.

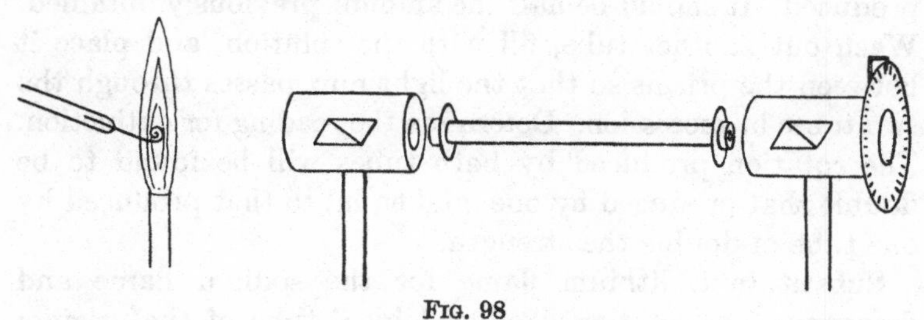

Fig. 98

The Nicol without the circular scale is to be placed between the flame and one end of the tube, *i.e.* used as the "polariser," and that with the index and circular scale graduated in degrees, placed between the other end of the tube and the eye, *i.e.* used as the "analyser," Figs. 97 and 98.

A thin sheet of yellow glass may be placed between the flame and polariser to cut off the blue light of the flame, and if necessary a lens may be used between the analyser and the eye.

Wash out the tube thoroughly, and after filling it with distilled water place it between the prisms and rotate the analyser till no light passes through the system. This will be the case when two lines similarly placed in the Nicol's and at right angles to the line of sight are also at right angles to each other. Read the circular scale on the analyser.

Repeat the observation several times, approaching the point of extinction from opposite sides each time, and take a mean of the results.

Dry the tube by passing through it a plug of cotton-wool, and then fill it with the salt or sugar solution, place it between the prisms, and determine as before the position of the analyser when no light passes. Repeat, and take the mean.

The difference between the two means is the rotation of sodium light produced by the solution.

To test the truth of the law expressed by the equation $R = a.l.\rho$, take 20 c.c. of the solution, and dilute to 40 c.c. Fill a tube with this solution, and determine the rotation produced. It should be half the amount previously obtained. Wash out another tube, fill with the solution, and place it between the prisms so that the light now passes through the two tubes in succession. Determine the reading for extinction. The rotation produced by both tubes will be found to be double that produced by one, and equal to that produced by one tube of double the strength.

Substitute a lithium flame for the sodium flame and determine the rotation produced by a tube of the original solution. It will be found less than that for sodium light.

Wash out the tubes thoroughly before putting away the apparatus.

Arrange your results as follows:

Date:

Rotation of plane of polarisation by sugar solution.

A. Sodium Light.

Reading for darkness, with water tube ... ... 167°·2
" " with original solution ... 156°·4
Rotation for original solution ... ... = 10°·8
Reading for darkness, with solution of half strength = 161°·7
Rotation for solution of half strength ... = 5°·5
Reading for darkness, with 2 tubes of solution of half strength ... ... = 156°·1
Rotation for double length of solution of half strength ... ... = 11°·1

B. Lithium Light as above.

SECTION XLVI

SACCHARIMETRY

Apparatus required: *Laurent or other polarimeter, sugar, and sodium flame.*

With the apparatus used in the previous exercise the eye has been called upon to judge the point at which the minimum light passed through the Nicol's prisms. As the estimation of the exact position for a minimum is difficult and the practical determination of the amount of sugar in a sample is made to depend on its power of rotating the plane of polarisation of light, several instruments have been devised for getting over the difficulty, one of these being the Laurent polarimeter provided. By their means it is possible to estimate the position of extinction to a small fraction of a degree.

The Laurent instrument is arranged as shewn in Figs. 99, 100.

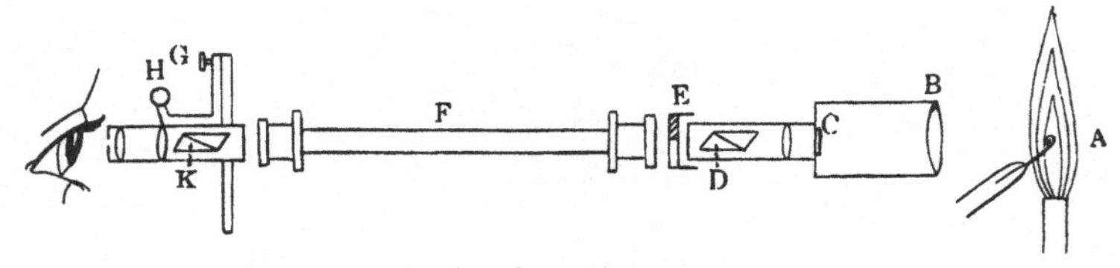

FIG. 99

The light from the Bunsen burner A, in which a bead of a mixture of equal parts of sodium diborate and common salt is placed, passes through the lens B, the small hole in the diaphragm C, which is covered by a thin plate of bichromate of potash to cut off all but the yellow light, and falls on the lens and Nicol prism D. The plane polarised light which emerges from the Nicol falls on a plate of quartz E, which covers half the field. The quartz plate is cut so that the optical axis is parallel to the edge which bisects the field. The plane polarised light falling on the plate is decomposed into two rays, one polarised in a plane parallel to the edge of the plate, the other in a plane perpendicular to this edge. The two rays traverse the plate with different velocities, and the thickness of the plate is so arranged that a difference of phase of half a wave length is produced. The effect of this is, that if the light passing through the uncovered half of the field is polarised, say in the direction CA inclined at an angle θ to CB the edge of the quartz plate, then that which has passed through the plate is polarised in the direction CA' such that $BCA' = BCA$. On looking through the analysing Nicol K of the eyepiece, the two halves of the field will appear unequally bright, unless the principal plane of the analysing Nicol makes equal angles with the directions CA, CA', *i.e.* is either parallel or perpendicular to CB. If it is parallel to CB the halves are equally bright, if perpendicular equally dark. The dark position is the one made use of, and the instrument is more sensitive the smaller θ is consistently with sufficient light passing through the apparatus. Generally θ does not exceed 2°.

A'B A

C

FIG. 99 *a*

Adjust the eyepiece till the dividing line between the fields is seen distinctly when one half of the field is dark and the other light.

The Nicol D is connected to a horizontal moveable arm, shewn in Fig. 100, and may be rotated within certain limits. In order to get the position of greatest sensitiveness, determine first the position of the arm when the two halves of the field are equal in brightness whatever be the position of the Nicol K. This should be done with an empty tube at F. Now rotate D through a small angle not exceeding 2° by means of the moveable arm. It will then be found that the two halves of the field are equally dark for a certain position of K, the left-hand half increasing, the other decreasing, in brightness if K is moved in one direction, and the right-hand half increasing, the other decreasing, if the rotation is in the opposite direction.

By means of the rotating screw G, adjust the vernier to read 0, and by means of the tangent screw H, rotate K till equality of fields is again produced. The instrument now reads 0 when no active material is present.

Insert at F a tube filled with distilled water, and determine the reading for equality of the fields, the adjustment being made alternately from opposite sides of the position of equality. This reading should still be 0 if the water was pure, and the glass ends of the tube unstrained.

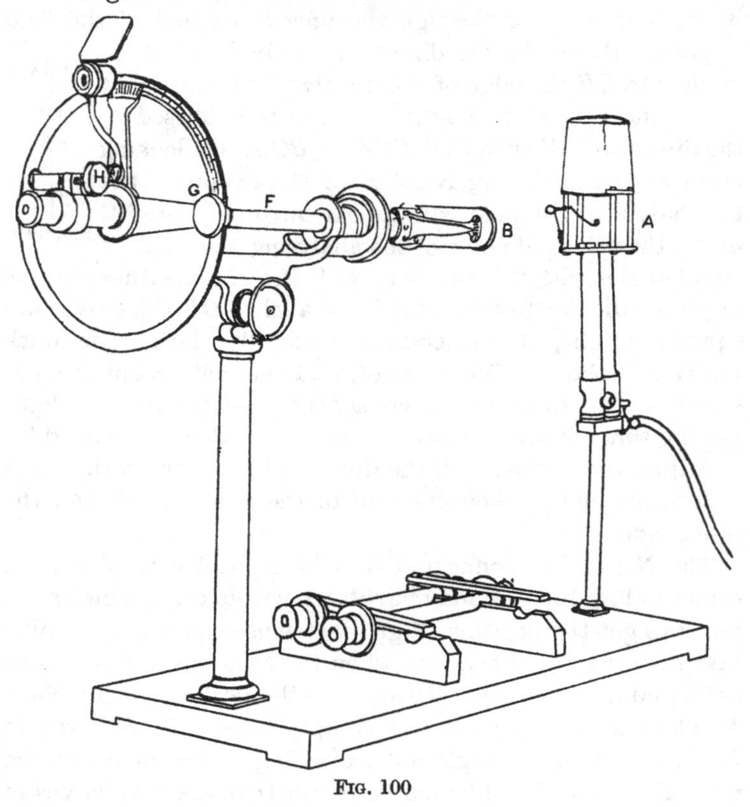

FIG. 100

Now insert a tube filled with a sugar solution containing 10 grams of sugar per 100 c.c. made and clarified according to the instructions, p. 228, and take readings several times as before.

The sample of sugar from which the solution has been derived may contain both cane sugar and invert sugar, and impurities which will be assumed to be inactive. The cane sugar rotates the plane of polarisation to the right, and the invert sugar to the left. The observed rotation will be due to the difference of these effects.

To determine the amount of each constituent present, we make use of the fact that when cane sugar is heated gently with acid it is converted into invert sugar, so that the whole of the sugar then present in the solution is invert sugar.

From the two observations of the rotation the amount of each constituent present can be calculated.

Take 20 c.c. of the original solution of sugar in a flask, add to it ·01 gram of strong hydrochloric acid per gram of sugar present, *i.e.* in the present case about 2 drops, add water till the volume is 22 c.c., and warm gently for 10 minutes, keeping the temperature about 70° C.

Cool the resulting invert sugar solution by placing the flask for 10 minutes in cold water, and after its temperature has fallen to about 20° C. note the temperature, insert the solution in the tube, and determine the rotation produced.

Let R_1 be the rotation observed with the original sugar solution, and let R_2 be that observed with the invert sugar, rotation to the right being considered positive and to the left negative. Let ρ_1 be the specific rotatory power of cane, ρ_2 that of invert sugar, and let a_1 grams of cane sugar and a_2 grams of invert sugar be present in 1 gram of the sample.

Then in the first experiment,

$$R_1 = \rho_1 . \frac{10a_1}{100} . 20 + \rho_2 \frac{10a_2}{100} . 20$$

$$= 2\rho_1 a_1 + 2\rho_2 a_2 \text{ ....(1)},$$

and in the second experiment,

$$R_2 = \rho_2 \left\{ \frac{10a_1}{100} . \frac{360}{342} + \frac{10a_2}{100} \right\} \frac{20}{22} . 20.$$

The factor 360/342 being due to the change from $C_{12}H_{22}O_{11}$ to $C_{12}H_{24}O_{12}$ in inversion of the sugar, and the factor 20/22 due to dilution with acid. Reducing we have

$$1 \cdot 1 R_2 = 2\rho_2 \{a_1\, 360/342 + a_2\} \text{ ....(2)}.$$

Subtracting (2) from (1) we have

$$R_1 - 1 \cdot 1 R_2 = (2\rho_1 - 2 \cdot 105\rho_2)\, a_1 .$$

Hence

$$a_1 = \frac{R_1 - 1{\cdot}1R_2}{2\rho_1 - 2{\cdot}105\rho_2}, \text{ and } a_2 = \frac{R_1/2 - \rho_1 a_1}{\rho_2} = \frac{1{\cdot}045R_2\rho_1 - R_1\rho_2}{2\rho_2\,({\cdot}95\rho_1 - \rho_2)}.$$

The quantity ρ_1 has been found to be nearly independent of temperature. Its value is $+ 6{\cdot}65$ degrees.

ρ_2 is negative and depends on the temperature t. Its value for sugar inverted in the way described is $- 2{\cdot}00 + {\cdot}031\,(t - 20)$ degrees.

Record observations and results as follows:

Date:

Solution containing 10 grams of sample per 100 c.c.

Reading with water in tube ...	...	0° 10′ to right.
Do. sugar solution 12° 44′, 12° 50′, 12° 50′, 12° 44′	mean,	12° 47′ ,, ,,
∴ Rotation at temp. 18° C. ...	...	12° 37′ ,, ,,
Reading with invert solution 3° 20′, 4° 0′, 3° 35′, 3° 45′	mean,	3° 40′ to left.
∴ Rotation at 18° 5′ C. ...	...	3° 50′ ,, ,,
Value of ρ_2 ,, ,,	...	$= - 2^\circ{\cdot}04$

$$R_1 = 12^\circ{\cdot}62, \qquad R_2 = - 3^\circ{\cdot}83.$$

$$\text{Hence}\quad a_1 = \frac{12{\cdot}62 - 1{\cdot}1\,(- 3{\cdot}83)}{13{\cdot}30 - 2{\cdot}105\,(- 2{\cdot}04)} = \frac{16{\cdot}83}{17{\cdot}59} = {\cdot}96,$$

$$a_2 = \frac{6{\cdot}31 - 6{\cdot}39}{- 2{\cdot}04} = \frac{{\cdot}08}{2} = {\cdot}04.$$

Or the sample contains 96 % of cane sugar and 4 % of invert sugar.

Wash out the tubes thoroughly with tap water before putting away the apparatus.

BOOK VI

MAGNETISM AND ELECTRICITY

SECTION XLVII

HORIZONTAL COMPONENTS OF MAGNETIC FIELDS

Apparatus required: *Mirror magnetometer, lamp and scale, bar magnet, vibration box, tangent galvanometer and milli-ammeter.*

(*a*) **Magnetometer Method.**

If M is the magnetic moment of a magnet and l half the distance between its poles, *i.e.* about $\frac{5}{12}$ the length of the magnet, the magnetic field I it produces at a point P distant d from its centre along its axis is along the axis and

$$I = M \cdot \frac{2d}{(d^2 - l^2)^2} \quad \text{.....................(1).}$$

If the point P is so far from the magnet that l is small compared to d, we have simply

$$I = M \cdot \frac{2}{d^3} \quad \text{.........................(2).}$$

At a point P distant d from the centre C of the magnet along a line through C perpendicular to the axis we have

$$I = \frac{M}{(d^2 + l^2)^{\frac{3}{2}}} \quad \text{.......................(3)}$$

and is parallel to the axis.

If P is so far from C that l is small compared to d, this becomes

$$I = \frac{M}{d^3} \quad \text{.........................(4).}$$

Hence the field at a distance d from the centre of a short magnet along the axis of the magnet is twice the field at the same distance along a line through the centre of the magnet perpendicular to the axis. If the law of force had been the inverse nth power instead of the inverse square, the former force would have been found to be n times the latter. Hence an experimental determination of the ratio of these forces will furnish a proof of the law of action of magnetic poles on each other.

If the axis of the magnet is at right angles to the magnetic meridian, and H is the horizontal component of the earth's magnetic field, the resultant of the two fields will make an angle θ with the meridian, where

$$\tan\theta = \frac{I}{H} \quad \text{.........................(5).}$$

The angle θ may be determined by placing a small magnetic needle at the point P, and observing the deflection produced when the magnet is placed in position.

The apparatus provided Fig. 101 consists of a small magnetic

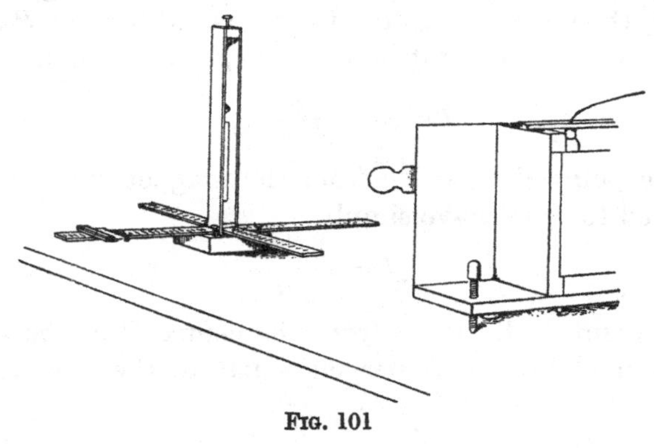

Fig. 101

needle to which a mirror is attached, suspended by a fine fibre, the torsion of which may be neglected. The centre of the needle is situated 3 mms. above the middle point of the upper of the two horizontal graduated scales placed at right angles to each other. One of the scales is placed in the magnetic meridian, *i.e.* parallel to the axis of a magnetic needle brought near it. For convenience this scale should be about 2 mms. below the other. The angle this direction makes with the walls of the room should be observed and compared with the angle the geographical meridian makes with the walls. The latter is shewn by a mark on the floor drawn along the shadow of a vertical bar of a south window cast by the sun at noon on April 15, June 15, August 31 or December 24. The rotation of the mirror is determined in the usual manner by the motion of the image of a cross wire formed after reflection by the mirror, on a scale placed parallel to the mirror when in its central position. The cross wire is attached to the scale and illuminated by a lamp. (For the method of obtaining the angles of rotation see Section XXXI.)

The position of the image of the cross wire on the scale is first read.

The deflecting magnet provided is then placed on the scale running east and west, say to the west of the needle, with its north pole towards the needle, the positions of the ends of the magnet are observed, and the scale reading of the cross wire is taken. The magnet is then reversed so that its south pole now points towards the needle, and the deflection again observed. It is then transferred to the east of the needle and the two deflections again observed.

To make the corresponding observations with the magnet north and south of the needle, the magnet is placed on a frame which slides along the scale in the magnetic meridian, the magnet itself being at right angles to the meridian. The height of the frame is such that the magnet itself will be at the same level as when placed on the other scale. The frame is moved along the scale till the centre of the magnetic axis is at the same distance from the centre of the needle as previously, the north pole of the magnet being say to the east. The deflection of the cross wire is observed, the magnet is reversed so that its north pole points to the west, and the observation repeated. The frame and magnet are then transferred to the other side of the needle and the observations repeated.

A set of observations should be taken for each of three distances of the magnet from the mirror as far apart as possible, and each set recorded as follows:

Date:

Determination of M/H at station B.

Inclination of magnetic meridian scale to normal to south wall of room ... ... 30° west.
Inclination of geographical meridian line to normal to south wall ... ... 12° ,,
Deviation of compass ... ... 18° ,,
Deflecting magnet A, length = 6·0 cms. $\therefore l = 2{\cdot}5$ cms.
Magnetometer B, distance of scale from mirror = 103 cms.

Position of Magnet			Magnetometer		mean	tan 2θ	tan θ
N. end cms.	S. end cms.	Middle cms.	Reading cms.	Deflection cms.			
Position of rest			50·0	zero			
25·0 W	31·0 W	28·0 W	68·5	18·5			
31·0 W	25·0 W	28·0 W	31·6	18·4			
31·0 E	25·0 E	28·0 E	69·3	19·3			
25·0 E	31·0 E	28·0 E	30·2	19·8	19·0	·184	·092

Similarly for the north and south positions.

For the east and west positions in the above example we have:

$$\frac{M}{H} = \frac{(d^2 - l^2)^2}{2d} \tan\theta = \frac{\{28^2 - (2{\cdot}5)^2\}^2}{56} \cdot 092 = 996.$$

Similarly for the other positions.

Find the quotient of tan θ in the east-west by that in the north-south positions and see how closely it approximates to 2.

In order to determine both M and H it is necessary to find the value of some combination of the two other than the quotient, and the most convenient combination to determine experimentally is the product MH, which is connected with the time of torsional oscillation τ of the magnet about a vertical axis through its centre of gravity, by the equation

$$\tau = 2\pi \sqrt{\frac{I}{MH}},$$

where I is the moment of inertia of the magnet about the axis of oscillation.

Hence

$$MH = \frac{4\pi^2 I}{\tau^2}.$$

To determine the product MH, place the magnet which has been used as the deflecting magnet in a light paper or silk stirrup supported by a long thin fibre the torsion of which can be neglected, so that it can oscillate in a horizontal plane about a vertical axis. To protect the magnet from air currents suspend it in a box. Observe the times at which one end of the magnet passes a fixed mark on the bottom of the box in one direction, for six consecutive passages, the arcs of vibration not exceeding 20°. Wait a time nearly equal to that between the first and last observations, and then take six more observations. Arrange as follows:

Passage No.	First Set		Passage No.	Second Set		Time of 10 oscillations
0	1 h. 3 m.	2 s.	10	1 h. 4 m.	11 s.	69 secs.
1		9	11		17	68 ,,
2		16	12		24	68 ,,
3		23	13		30	67 ,,
4		29	14		38	69 ,,
5		36	15		43	67 ,,

Mean ... ... = 68 secs.
Time of one oscillation τ ... ... = 6·8 ,,

Determine the moment of inertia of the bar magnet by weighing it to ·01 gram, and measuring its length and breadth. The moment of inertia I, if m is the mass of the magnet, $2a$ its length, $2b$ its breadth, is given by the equation

$$I = m.(a^2 + b^2)/3.$$

In order to determine whether a magnet is weak or strong, it is advisable also to calculate the magnetic moment per unit mass.

Arrange observations and calculations as follows:

Mass of magnet A = 11·01 grams.

$$\left.\begin{array}{ll} a = 3{\cdot}0 \text{ cms.} & a^2 = 9{\cdot}0 \\ b = {\cdot}23 \text{ ,,} & b^2 = {\cdot}05 \end{array}\right\} a^2 + b^2 = 9{\cdot}05,$$

$$\therefore I = 11{\cdot}01 \times 3{\cdot}02 = 33{\cdot}2$$

and $$MH = 4 \times 9{\cdot}87 \times 33{\cdot}2/46{\cdot}2 = 28{\cdot}4.$$

Having obtained M/H and MH, M and H are determined as follows:

$$M = \sqrt{(M/H)\,(MH)} = \sqrt{996 \times 28{\cdot}4} = \sqrt{28300} = 168.$$

$$H = \sqrt{MH/(M/H)} = \sqrt{28{\cdot}4/996} = \sqrt{{\cdot}0285} = {\cdot}169.$$

Magnetic moment of magnet per gram $= 168/11{\cdot}01 = 15{\cdot}2$.

(*b*) **Galvanometer Method.**

The considerable time required for a determination of H by the magnetometer method has led to the introduction of methods depending on the comparison of H with the magnetic field due to the flow of a measured electric current round a coil of known dimensions. The various methods suggested differ in the way in which the comparison is made. One of the best of these not requiring special apparatus is to use a double coil galvanometer (Helmholtz) with its axis nearly in the magnetic meridian and to send through it a current measured on a standard ammeter or better by the standard resistance and standard cell method of Section LXIII, which produces a field F nearly equal and opposite to H. If the current is adjusted so that when the axis of the coil makes a small angle α with the magnetic meridian the resultant field is at right angles to the axis of the coil,

$$H = F/\cos \alpha.$$

Determine H at the previous place of observation by this method.

Set up a double coil tangent galvanometer on a base which can be rotated 5° or 10° in either direction. See that the zeros of the scale are in a line perpendicular to the planes of the coils.

Connect the galvanometer through an adjustable resistance and a milliammeter standardised by the method of Section LXIII to one or two storage cells.

Verify that when no current passes the readings are zero. Rotate the galvanometer till the reading is 5° west at north end. Send a small current through the galvanometer in such a direction that the deflection is increased, and increase the current till the deflection becomes 90° west at north end. Read the current.

Now rotate the galvanometer to the other side of the meridian till the deflection becomes 90° east at north end. Read the current and stop it. Read the position of the needle at no current.

Measure the mean radius r of the coils and their distance apart. These three should be equal, and the magnetic field F at the point on the axis half-way between the coils is given by $F = \frac{16\pi}{5\sqrt{5}} \cdot \frac{n}{r} \frac{A}{10}$, where A is the current in amperes, and n the total number of turns.

Record as follows:

Date:

Double Coil Galvanometer C.

Number of turns, 72.

Mean radius of coils, 9·40, distance apart 9·41 cms.

Deflection without current	+ 5°	with ·050 amp.	+ 90°.
,, ,, ,,	− 5°·2	,, ·0503 ,,	− 90°.
Mean	5°·1	·0502	

$$H = F \sec \alpha = \cdot 1724 \times 1 \cdot 004 = \cdot 173.$$

16–2

SECTION XLVIII

MAGNETIC DIP

Apparatus required: *Dip circle instrument, earth inductor coil and low resistance galvanometer.*

(*a*) **Dip Circle Method.**

A dip circle instrument consists of a circular table supported on three levelling screws by means of which it can be set horizontal. This table supports in turn a concentric circular disc to which is attached a vertical circular ring and a spirit level. The disc and ring are graduated in degrees and sometimes in minutes of arc, the disc or azimuth circle from 0° to 90° in each quadrant and the ring or dip circle from 0° to 90° in each direction from a horizontal diameter. Within the dip circle a pointed magnetic needle with an axle through its centre perpendicular to its plane can be placed, the axle being supported either on two horizontal agate knife edges on which it can roll, or in the case of marine instruments in jewels into which its conical ends project. In the former arrangement the axle can be raised from the agates by means of two Y-shaped supports. To protect the needle from air currents the dip circle is generally enclosed in a box with glass sides.

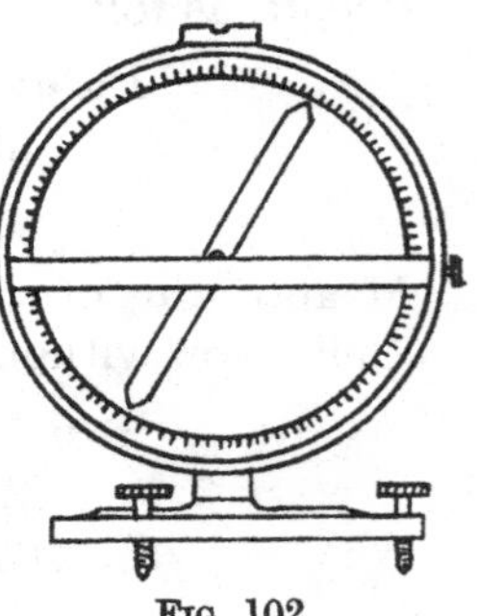

Fig. 102

The position of the ends of the needle on the dip circle may be read either directly or by the help of microscopes.

The side of the ring on which the scale is engraved and that side of the needle on which a line or letter is engraved will be called the face in each case.

To set up the instrument so that the surface of the base plate on which the vertical circle turns is horizontal, turn the dip circle till its plane is parallel to the line joining two of the levelling screws of the base which have been put down roughly north and south of each other. Adjust one of the two screws till the bubble is in the middle of the level, then turn the dip circle front to back. If the bubble has moved from the centre bring it half-way back by turning one of the two levelling screws. Turn the dip circle back to the original position. If the bubble remains in the same position on its scale when the dip circle is turned front to back the surface of the base plate on which the vertical circle turns is horizontal in a north and south direction. Now turn the dip circle through 90° and adjust the third levelling screw till the bubble remains stationary on its scale during the rotation. It will then remain stationary however the dip circle is rotated, and the surface on which that circle turns is horizontal.

The needle should now be removed from its box with tweezers, the axle cleaned by being thrust into a piece of pith or soft cork, the agate knife edges and the axle dusted with a camel hair brush and finally the needle placed on the Y stirrups with its face to the face of the dip circle. After the box is closed the stirrups should be lowered by the screw outside, and the axle allowed to rest on the agates. Rotate the dip circle on the azimuth circle till the upper end of the needle reads 90° on the dip circle. Read the azimuth circle. Rotate the dip circle till the lower end of the needle reads 90°. Read the azimuth circle. The two readings should be nearly alike. Turn the dip circle front to back and repeat the two settings. Turn the needle front to back on the agates, using the tweezers for the purpose, and repeat the four settings. The mean of the eight readings gives the best azimuth setting for the axle of the needle to be magnetic north and south. Rotate the dip circle through 90° in azimuth, so that the faces of dip circle and needle are to the magnetic east, the axle is then magnetic east and west and the needle moves in the magnetic meridian.

As the oscillations of the needle die down read three consecutive turning points first for the upper then for the lower end, and calculate the position of rest of each end as in the case of the balance (p. 33).

Turn the circle through 180° so that the face is west and repeat the observations. Turn the needle front to back on its bearings so that the front is now to the east, and repeat the observations. Finally bring the circle back 180° so that its face is again east and repeat the readings.

Remove the needle, place it in the wooden frame provided, cover the axle and reverse its magnetism by stroking it several times gently with opposite poles of the two bar magnets provided, one being held in each hand and the stroke of each magnet being away from the axle. Place the needle on its stirrups with its face to the face of the circle and repeat the whole of the observations.

The observations taken may be summarised as follows:

$$\text{Circle} \begin{Bmatrix} \text{east} \\ \text{west} \end{Bmatrix}, \text{ Needle} \begin{Bmatrix} \text{east} \\ \text{west} \end{Bmatrix}, \text{ Magnetisation} \begin{Bmatrix} \text{direct} \\ \text{reverse} \end{Bmatrix},$$

the first four alternatives referring to the faces of circle and needle.

Tabulate the observations as follows:

Date:

Dip Instrument *A*.

Azimuth readings with ends of needle at 90°:

21° 4′, 21° 0′ 21° 1′, 21° 3′ &c. Mean 21° 2′.

Circle rotated through 90° from mean reading.

	Top	Bottom	Mean
Ce Ne Md	67° 21′	67° 20′	67° 20·5′
Cw Nw Md	67 18	67 16	17
Cw Ne Md	67 19	67 16	17·5
Ce Nw Md	67 18	67 20	19
		Mean	67° 18·5′

Similarly for Mr.

The mean of the two means is taken as the final result.

(*b*) **Earth Inductor Method.**

It has been found that the dip can be more accurately measured by means of a coil rotating about an axis in its own plane in the earth's field. An alternating electromotive force is produced in the coil which only becomes zero when the axis of rotation coincides with the direction of the earth's field. The absence of electromotive force can be observed by means of a telephone or if the coil is provided with a commutator by means of a galvanometer.

By means of a compass needle place the earth inductor so that its axis of rotation is in the magnetic meridian and inclined to the horizontal at an angle of 60°.

See that the brushes make good contact with the commutator and that they change over from one segment to the other when the axis of symmetry of the coil is in the vertical plane through the axis of rotation.

Connect the brushes to a low resistance galvanometer of 10 to 20 ohms, and rotate the coil by hand at a steady speed of 1 to 2 revolutions per second. Observe the deflection of the galvanometer and alter the inclination of the axis of the coil keeping it in the magnetic meridian till the deflection is reduced to zero.

If the brushes can be rotated 90° it can in the same way be determined whether the axis of rotation of the coil is in the magnetic meridian. When the axis of rotation is along the field there should be no deflection produced whatever the position of the brushes.

Measure the length of the axle of the coil and the heights above the table (assumed horizontal) of the two centre points at the ends of the axle. The quotient of the difference of the heights by the length of the axle is the sine of the angle of dip.

Record as follows:

Date:

Earth Inductor *B*. Galvanometer 15.

Inclination of magnetic meridian to edge of bench	43°
Length of axle	41·51 cms.
Height of centre of upper end	48·71 ,,
,, ,, lower ,, ...	10·45 ,,
Difference	38·26 ,,

$$\sin(\text{dip}) = \frac{38\cdot26}{41\cdot51} = \cdot919,$$

$$\text{Dip} = 67^\circ\ 48'.$$

SECTION XLIX

MAGNETISATION CURVES

Apparatus required: *Reflecting Magnetometer and scale, magnetising coil, thin iron rod, standardising coil, ammeter, storage cells, adjustable resistance and reversing switch.*

In this method a thin rod of the material to be tested is placed in a magnetising coil, the field of which can be varied, and the strength of the pole produced at the end of the rod is measured by a magnetometer.

Arrange the long coil B (Fig. 103) vertical, clamping it to the edge of the bench.

Place the magnetometer M about 20 cms. magnetic west from the axis of the coil, the scale S about 70 cms. west of the magnetometer, and the compensating and standardising coil C between the magnetometer and scale.

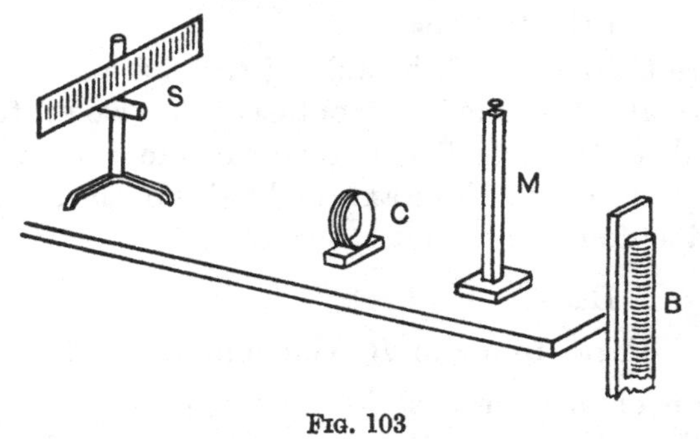

Fig. 103

The coil B consists of 5 layers of No. 22 wire, 768 turns per layer. The terminals connect to 1, 1 and 3 layers respectively. The magnetic field F along the axis of the solenoid well away from its ends is given by $F = 4\pi N_1 A/10$, where N_1 is the number of turns per cm. and A is the current round them in amperes.

Connect to the terminals of one layer a small storage cell with a resistance of about 80 ohms in series so as to annul the vertical field of the earth in the coil. As seen from the top of the coil a current passing from the top to the bottom terminal flows in the counter-clockwise direction, and such a current will tend to neutralise the vertical field. From the number of turns of the coil calculate the current necessary to exactly neutralise the field, and place sufficient resistance in circuit to get this current from the storage cell.

Adjust the magnetometer so that it swings freely and gives a clear spot of light on the scale. Connect up as shewn in Fig. 103 *a* with a reversing key *R*, a moving coil ammeter, and a slide rheostat in circuit. Send about 1 ampere through the solenoid and notice the deflection of the magnetometer. Place the compensating coil *C* in circuit, keeping the coil as far as possible from the magnetometer. Notice whether, on again sending one ampere through the solenoid and cell, the deflection of the magnetometer is greater or less than it was previously; if greater, reverse the connections of the coil, and the deflection will be less. Bring the coil nearer to the magnetometer until the deflection is reduced to zero.

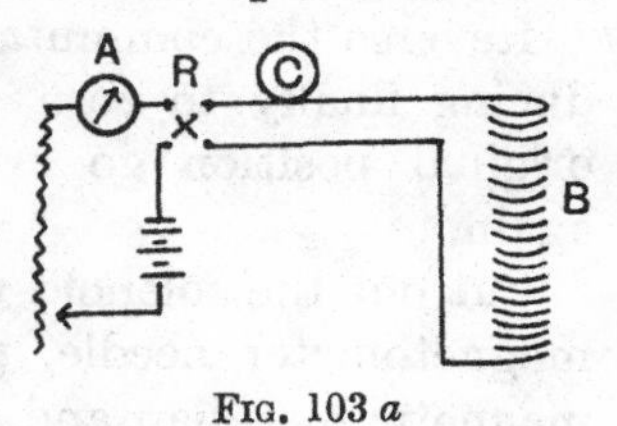

Fig. 103 *a*

Reverse the current and see that there is still no deflection. If there is, some part of the circuit is too near the magnetometer and must be removed to a greater distance.

Stop the current and place the iron wire whose magnetisation curve is to be obtained, with its axis in the axis of the solenoid, and a point about $\frac{1}{12}$th of its length from its top end in the same horizontal plane as the magnet of the magnetometer.

In general the wire will be found to deflect the magnetometer a little as it will be slightly magnetic. Magnetise it by sending about 1 ampere through the solenoid. Notice the deflection and move the solenoid from the magnetometer until it is about half across the scale. To demagnetise the wire cut the ammeter out of circuit and reduce the current to zero slowly by putting in resistance, reversing the current once or twice a second during the whole time the current is being reduced. At the end of the operation the deflection of the magnetometer should be zero or very small and the experiment can be commenced.

Send currents of about ·1, ·2, ·3, ·4, &c., ampere, up to about 2 amperes, through the solenoid, observing the current and the deflection of the magnetometer in each case.

Then diminish the current in steps of ·1 ampere to zero, observing again the currents and the deflections.

Reverse the commutator and send currents as before, reducing finally to zero, and then with commutator in the original position go over the first series of observations again.

Cut out the solenoid with its centre magnetic west of the magnetometer needle, place the compensating coil in the magnetic meridian and send sufficient current through it to give a deflection about equal to $\frac{3}{4}$ of the largest obtained during the observations. Observe the current and the deflection.

Measure the distance of the centre of the coil from the centre of the magnetometer and the radius of the coil a.

If F' is the field at the magnet due to the current A in the coil,

$$F = \frac{2n\pi C}{a}\sin^3\phi$$

$$= \frac{2n\pi A}{10a}\sin^3\phi.$$

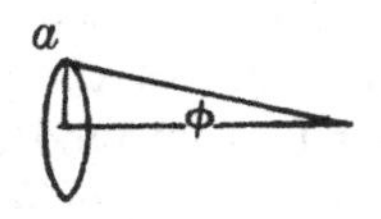

FIG. 104

FIG. 104 *a*

Calculate the field f corresponding to a deflection of 1 cm. Measure the distance of the needle from the axis of the solenoid, and the length of the wire. Then if I is the magnetisation of the wire and S is its cross section and $2l$ its length, the field at the magnet due to the wire

$$= \frac{IS}{d^2}\left(1 - \left(\frac{d}{d'}\right)^3\right),$$

where d is the distance of the upper pole of the wire, and d'

of the lower pole from the centre of the magnetometer needle. Hence $f \times$ deflection

$$= \frac{IS}{d^2}\left(1 - \frac{d^3}{d'^3}\right)$$

or

$$I = \frac{d^2}{S} \frac{1}{1 - \frac{d^3}{d'^3}} \cdot f \times \text{deflection.}$$

Calculate I for each deflection of the magnetometer. From I calculate B, the magnetic flux per square centimetre, using the relation $B = F + 4\pi I$. Tabulate the results as follows:

Date:

Magnetising coil 1.

$d = 20{\cdot}2$ cms.

$d' = 85{\cdot}1$,,

Wire iron 80 cms. long, ·203 cm. diameter.

A	F	Deflection	I	B
+ ·1	·240	1·25 cms.	17·1	216
+ ·2	·480	2·56	34·9	440
	&c.		&c.	

Plot curves shewing the relation between I and H, and between B and H, H being abscissae in each case.

SECTION L

THE WATER VOLTAMETER

Apparatus required: *Water Voltameter, two storage cells, tangent galvanometer (with a constant of* 1 *or* 2), *ammeters for* 1–2 *amperes to be standardised, reversing key and resistance coils.*

When an electric current decomposes a liquid through which it is sent, the liquid is called an "electrolyte," and the process of decomposition is called "electrolysis." The laws of electrolysis state that the weight of an element set free from a compound by electrolysis is proportional to the current, the time, the atomic weight, and inversely proportional to the valency of the element. One ampere flowing for one second liberates, *e.g.*, ·0011183 gram of silver from *any* silver salt. The amount of an element liberated by a current in a given time may therefore be used to determine the magnitude of the current and hence to standardise a current-measuring instrument.

In the present exercise a tangent galvanometer and several other current-measuring instruments in series are to be standardised by means of a water voltameter. There are two ways of using a water voltameter, according as the hydrogen alone is measured, or both oxygen and hydrogen are collected together. In the latter case we gain by having a greater volume of gas to measure and being able therefore to reduce the time of the experiment, but there is some danger of error owing to the formation of ozone and its absorption in the water. It has however been shown by Kohlrausch that if currents of over one ampere are to be measured with an accuracy sufficient for commercial purposes, a voltameter in which both gases are collected together will answer the purpose. For smaller currents the hydrogen only should be measured. The difficulty may also be overcome by using a solution of sodium hydrate of density about 1·06 with pure nickel electrodes.

The volume of mixed gas produced by the passage of an ampere for a second is ·1734 c.c. at normal temperature and pressure. The gas is collected in a graduated tube and its volume measured. To reduce the observed volume to the normal conditions, the temperature of the gas must be measured by a thermometer placed, if possible, inside the voltameter tube. The pressure is obtained by reading the barometer, by applying if necessary a correction for a difference of pressure inside and outside the voltameter tube, and by *deducting the pressure of aqueous vapour in the tube*. As pure water offers a high resistance to the electric current a little sulphuric acid is added to it. If the reductions are to be made to the highest possible accuracy, account must be taken of the fact that the pressure of saturated vapour of water is less over a solution of sulphuric acid then over pure water. But in the present exercise it is assumed that the acid added to the water is not sufficient to produce a sensible effect.

One form of voltameter (Fig. 105) consists of two portions,

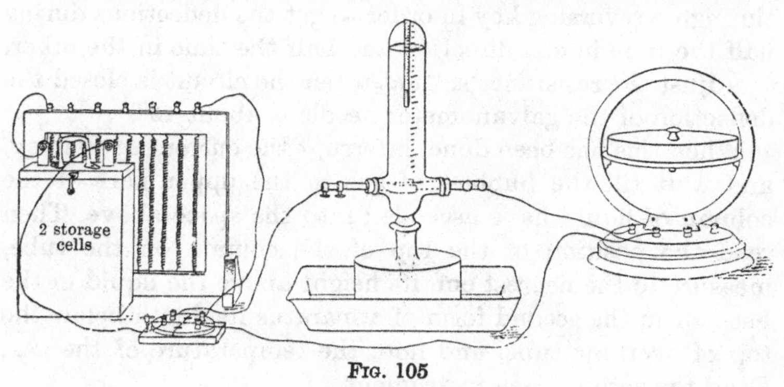

FIG. 105

a lower reservoir and a graduated tube which fits into it by means of a ground joint. The lower reservoir communicates with the atmosphere through an opening which can be closed by a glass stopper. *During electrolysis this stopper must be removed.* Pour water acidulated with about 15 % of dilute sulphuric acid into the reservoir so as to render it about three-quarters full. Insert the tube and stopper and tilt the voltameter so that the liquid runs into and completely fills the tube. Place the instrument in a tray and remove the stopper.

Another form consists (Fig. 105 *a*) of a vertical tube graduated in c.c.s., from the bottom of which a narrower tube leads to the top of a tube similar to the first. By tilting the apparatus to the left through about 90°, the acidulated water can be made to fill both the graduated tube and the narrow tube. As gas is generated in the graduated tube the water expelled runs over from the top of the narrow tube into the second wide tube, and its level in the narrow tube remains constant.

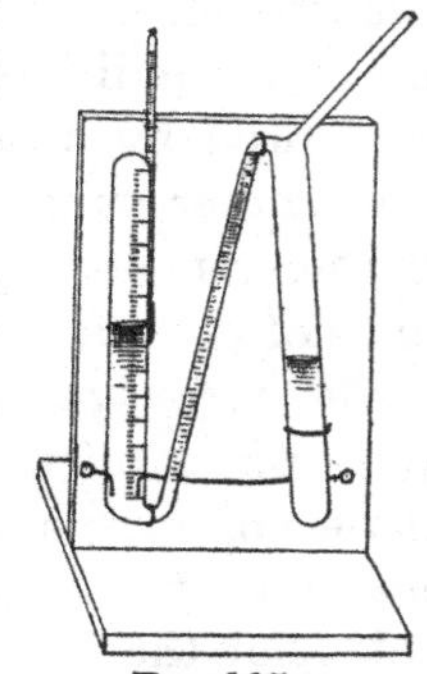

FIG. 105 *a*

Connect the instruments and voltameter in series with a resistance and two storage cells or other battery having an electromotive force of 3 or 4 volts and a low internal resistance. If any of the instruments read on both sides of zero they should be connected to the circuit through a reversing key in order to get the deflections during half the time in one direction and half the time in the other.

Adjust the resistance so that when the circuit is closed the deflection of the galvanometer needle is about 45°.

When this has been done, interrupt the current at the key, and wait till the bubbles of gas in the upper part of the column of liquid have ascended into the space above. Then read the position of the top of the column on the tube, measure to the nearest cm. its height above the liquid in the base, or in the second form of apparatus its depth below the top of overflow tube, and note the temperature of the gas. Read the zero of each instrument.

Now make the circuit at the key, noting the time, and at the end of a minute read the instruments, repeating the reading every two minutes till the tube is about $\frac{1}{3}$ full of gas. One minute after taking the last reading, break the circuit, noting the time. After allowing the bubbles of gas in the column of liquid to ascend, read the position of the top of the column and measure as before the height of the liquid column, and the temperature of the gas.

At a given instant make the circuit again with the reversing key so arranged that the current passes in the opposite direction through the centre zero instruments for an equal time. Read the instruments every two minutes as before, and the final position of the level of the liquid.

Read the barometer.

Calculate for each of the three observations of the column of liquid, the volume of gas reduced to normal temperature and pressure, and subtract the first from the second and the second from the third. The differences are the volumes of gas produced in the observed times by the passage of the current. Divide these by the times in seconds to get the volume per second, and then again by ·1734 to get the mean current in amperes. Compare the currents read on the instruments with this.

Take the mean of the tangents of the galvanometer deflections and the means of the readings of the other instruments for each period. Then if d be the mean tangent, and K the constant of the galvanometer, the average current passing is Kd amperes, hence K is determined.

Calculate K from the observation in this way, arranging the work as shewn below.

The two sets of observations are reduced separately so as to afford a check on the calculations and a test of the consistency of the results.

Remove the tangent galvanometer, and substitute for it one of the boxes used in measuring Magnetic Fields (Section XLVII). Determine the time of oscillation of the needle. Determine also the time of oscillation of the needle when placed at some point of the Laboratory at which the earth's horizontal magnetic force H is known, and from the observations calculate the value of H at the place where the galvanometer stood.

From the value of H and the radius and number of turns of the coil calculate the constant of the galvanometer, and compare the calculated value with the result of the experiment.

Arrange your observations as follows:

Date:

Voltameter *A*. Milliammeters 3 and 5.

Tangent Galvanometer *B*.

Height of Barometer = 75·67 cms.

Voltameter observations	I	II	III
Reading of voltameter tube*, c.c. ...	16·5	133·5	253·8
Height of column of water, cms. ...	35	26	15
„ „ equivalent mercury column	2·58	1·91	1·11
Pressure of gas, cms.	73·09	73·76	74·56
Temperature (Centigrade)	18°·8	18°·7	18°·5
Corresponding vapour pressure, cms.	1·61	1·60	1·58
Pressure of dry gas, cms.	71·48	72·16	72·98
Volume of gas reduced to 76 cms. and 0° C., c.c.	14·5	118·6	228·7

Current started at 11 h. 57 m. 12 h. 13 m.

„ stopped at 12 h. 07 m. 12 h. 23 m.

Volume generated	Time	Vol. per sec.	Mean amperes
118·6 − 14·5 = 104·1	600 secs.	·1735	1·000
228·7 − 118·6 = 110·1	600 „	·1835	1·058

Readings of milliammeters.

Time	No. 3	No. 5
11 h. 57·5	1·005	1·012
59	1·006	1·010
12 1	·998	1·002
12 3	·996	1·000
5	·990	·997
Means	·997	1·004
Errors	+ ·032	+ ·025

Similarly for the second set.

* If the tube is not divided into cubic centimetres, the value of the divisions ought to be ascertained by experiment.

Galvanometer Observations.

The readings of the ends of the pointer, marked *A* and *B*, are taken as positive when the deflections are counter-clockwise as measured from the zero divisions of the scale.

Time	Readings of Pointer			Deflections	Tangents
	End A	End B	Mean		
11 h. 57 m.	+ 0°·0	+ 0°·2	+ 0°·1		
11 58	+ 46·6	+ 46·6	+ 46·6	46·5	1·054
12 00	46·2	46·3	46·2	46·1	1·041
02	45·5	45·7	45·6	45·5	1·018
04	45·5	45·7	45·6	45·5	1·018
06	45·3	45·5	45·4	45·3	1·010
08	+ 0·0	+ 0·2	+ 0·1		
				Mean =	1·028

$$\therefore K = \frac{1\cdot000}{1\cdot028} = \cdot974.$$

Similarly for the observations during the second half of the experiment.

Constant of Galvanometer by calculation.

Time of oscillation at place of observation ... = 7·12 secs.
,, ,, standard position ... = 7·16 ,,
Value of *H* ,, ,, ,, ... = ·171 ,,
Value of *H* at place of observation ... ... = ·173 ,,
Mean radius of Coil of Galvanometer ... = 10·6 cms.
Number of turns ... ... = 3

$$\therefore \text{Constant} = \frac{5rH}{n\pi} \quad \dots \quad \dots \quad \dots \quad \dots = \cdot97$$

SECTION LI

THE COPPER VOLTAMETER

Apparatus required: *Current-measuring instruments to be standardised (one or two amperes suitable); storage cells, copper depositing cell, copper electrodes, plug-key, resistance coils, accurate balance.*

In the previous exercise the amount of electrolysis produced by the passage of a current was estimated by the volume of gas generated. Since volumes cannot be directly determined to the same degree of accuracy as masses can be weighed, more accurate results are obtained by the use of an electrolyte of which one of the products of decomposition is a solid which can be weighed. Silver and copper salts are found to be suitable for the purpose, as secondary reactions may be more completely avoided than in the case of other metals. For the most accurate work the silver voltameter is used, but its manipulation requiring great care, and the materials being expensive, copper may be substituted, and with proper precautions an accuracy of one part in a thousand may be attained.

In the present exercise a solution of copper sulphate in water will be used, and a current meter will be standardised by the weight of copper deposited. The example is worked out on the supposition that the instrument to be standardised is a Kelvin Current Balance, but any other instrument, such as that shewn in Fig. 106, may be standardised at the same time if placed in circuit.

Prepare a 20 % solution of copper sulphate, by adding 120 grams of copper sulphate crystals to 480 grams of water. When the crystals have dissolved, filter into the beaker provided and add one or two c.c. of strong sulphuric acid.

Clean the three copper plates provided, two of which are to serve as anodes and the third as cathode, with sandpaper.

Dip the one which is to serve as cathode into dilute nitric acid for about three minutes, then remove it and place the three in dilute sulphuric acid for about three minutes, finally wash under a stream of tap water.

Place the first two plates in the side clips A of the stand (Fig. 106) provided, and the third plate in the centre clip.

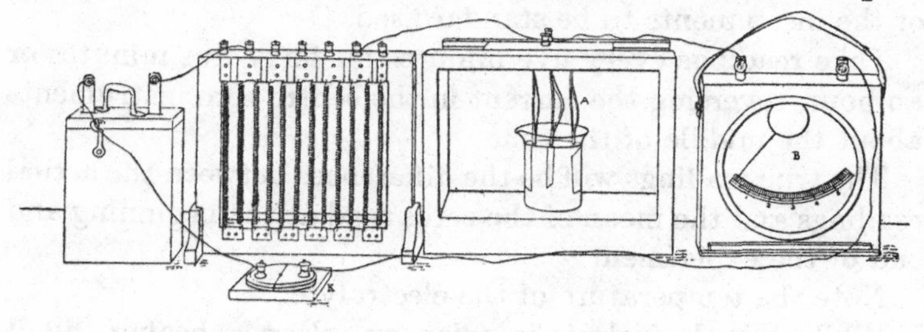

FIG. 106

Join up the copper voltameter, a plug-key K and an adjustable resistance to the instruments to be standardised, through two storage cells or other battery of small resistance giving an electromotive force of about 4 volts. Be careful to connect the terminals of the battery in such a way that the current passes through the copper sulphate solution from the outside plates (anodes) to the central plate (cathode) of the voltameter. If any of the instruments to be standardised are centre zero instruments, a reversing key should be inserted so that the direction of the current can be changed rapidly in them *without being reversed in the rest of the circuit*; and that change should be made in the middle of the experiment.

Adjust the resistances until the current is of convenient amount to be measured, then take out the plug-key K.

Raise the plates out of the solution, take the copper cathode from its clip, wash it in clean water, and dry first in a sheet of filter-paper, then before a fire, heating coil, or gas flame, taking care not to heat the plate appreciably. When the plate is dry and at the temperature of the room weigh it to ·1 milligram.

As the absolute weight is required, the plate must be weighed on both sides of the balance according to the method described on p. 46, and trustworthy weights only must be used.

Read the zero of each instrument.

Replace the cathode and lower the copper plates into the solution, make the circuit at K at a time to be noted in minutes and seconds, and as soon as possible take readings of the instruments to be standardised.

Take readings every five minutes for forty-five minutes or an hour, reversing the current in the centre zero instruments about the middle of the time.

The true readings will be the differences between the actual readings and the mean of the zeros read at the beginning and end of the experiment.

Note the temperature of the electrolyte.

While the electrolysis is going on, clean a beaker, fill it with clean water, and add a few drops of sulphuric acid.

At the end of the above period break the circuit at K, noting the time accurately, and read the zero of the instrument.

Raise the electrodes, remove the cathode and dip it as quickly as possible into the prepared acidulated water, then hold it under a gentle stream of water from the tap for a minute. Now dry it as before, first in a pad of clean filter-paper, then before a fire, heating coil or flame, and after cooling weigh again to ·1 milligram.

The relation between the current C, the electrochemical equivalent of copper z, and the weight W deposited in a given time t is

$$W = C.z.t, \therefore C = \frac{W}{z.t}.$$

In the absence of all secondary reactions the value of z should be constant. But owing probably to the presence of dissolved oxygen, the amount of copper deposited is not strictly proportional to the current and depends to a small extent on the temperature. These effects can be taken into account by making z depend on the current density at the cathode and on the temperature. Thomas Gray, who has carefully investigated the amounts of copper deposited under different conditions, has given the following values for the electrochemical equivalent of copper.

Area of cathode in square centimetres per ampere of current	Temperature 12°	Temperature 23°
50	·0003288	·0003286
100	·0003285	·0003283
150	·0003282	·0003280
200	·0003279	·0003277

The results of this table may be expressed by the equation

$$\cdot 0003288 - \cdot 0000003\,\frac{A-50}{50} - \cdot 0000002\,\frac{t-12}{11},$$

where A is the area of the cathode surface per ampere, and t the temperature. The value of A is found by calculating the current in the first instance approximately, using $\frac{1}{3000}$ as the equivalent.

Measure the total area of the two sides of the cathode, and obtain an approximate value of the area per ampere. From this and the temperature during the experiment find from the table the equivalent which applies to your experiment.

Record and reduce the observations as follows:

Date:

Standardisation of Kelvin Balance A and Ammeters 1 and 4.

Weight of cathode before electrolysis ... = 20·3798 grams
" " " after " ... = 21·5786 "
Amount of copper deposited ... ... = 1·1988 "
Duration of experiment = 1 h. 0 m. 25 s. = 3625 seconds
Current approximately $\frac{1\cdot 2 \times 3000}{3600}$ ... = 1·0 ampere
Total area of cathode ... ... = 66 sq. cms.
Area of cathode per ampere of current... = 66
Temperature ... ... = 18° C.
Equivalent of copper ... ... = ·0003286
Current = $\frac{1\cdot 1988}{\cdot 0003286 \times 3625}$ ... ... = 1·0064 amperes

Time	Reading of Kelvin balance	True reading	Square root	Ammeters	
				1	4
Zeros	·2			·002	·010
11 h. 0 m. on	409·6	409·4	20·23	1·021	1·062
5	406·6	406·4	20·16	1·019	1·060
10	406·4	406·2	20·15	1·019	1·059
15	405·4	405·2	20·13	1·018	1·059
⋮					
50	405·4	405·2	20·13	—	—
55	405·6	406·4	20·16	—	—
12 h. 0 m. 25 s. off	406·4	406·2	20·15	—	—
Zeros	·2			·002	·010
Means corrected for zero errors =			20·14	1·014	1·047

$$\text{Constant of balance} = \frac{\text{Mean current}}{\text{Mean square root of deflection}} = \frac{1{\cdot}0064}{20{\cdot}14}$$

$$= {\cdot}04997$$

Constant given by instrument maker ... ... $= {\cdot}05000$

Thus currents measured by the instrument using the constant ·05 would agree to within one part in a thousand with their value as determined by copper electrolysis. The constant supplied by the maker of the Kelvin balance is equal to the mean current divided by twice the square root of the reading; *i.e.* ·025 in the case of the above instrument. A table of doubled squared roots is provided with the instrument, and the last column in the above table may be replaced by one in which the double square roots are entered. But the method here adopted is applicable to all instruments of the dynamometer type and the square roots are easily found in Barlow's Tables*. If an instrument is of the tangent galvanometer type the square roots must be replaced by the tangents of the angles of deflection.

* If the square root of *e.g.* 408·4 is taken from Barlow's Tables, interpolation would seem necessary between the values of the roots given for 408 and 409, but this may be avoided by finding that number in the neighbourhood of 2000, the square of which has for its first four significant figures 4084. The position of the decimal point is obvious.

SECTION LII

ADJUSTMENT AND STANDARDISATION OF MIRROR GALVANOMETERS

Apparatus required: *Mirror galvanometer and scale, watch, megohm and Daniell cell.*

When the deflection of the moving magnet or coil of a galvanometer is to be measured with accuracy a mirror is attached to the moving part and the rotation is measured by the motion of a beam of light reflected from the mirror on to a scale parallel to the face of the mirror in the way explained in Section XXXI. Several small magnets are often used instead of a single one in order to increase the magnetic moment and so decrease the time of oscillation, and these are either attached directly to the back of the mirror by means of a little shellac, or the mirror is placed outside the coil and attached to a thin wire which passes through the coils and carries the magnets at its end. The suspended system must be free to move round a vertical axis, and is for this purpose attached to a quartz or silk fibre just strong enough to carry its weight and as free from torsion as possible.

In the moving coil instrument the suspension is a thin phosphor bronze strip which serves to convey the current to the coil. The other end of the coil is attached to a spiral of the same strip.

The deflection of the moving magnet of a galvanometer being increased by a diminution of the fixed magnetic field at the centre of the coil, we can render a galvanometer more sensitive by placing a permanent magnet near the instrument in such a position that it produces together with the earth's field a resultant field of the desired strength. Such a magnet is also necessary when the galvanometer has to be set up with the plane of the coil not in the magnetic meridian. For a magnet of known magnetic moment we find by calculation a position such that it will nearly neutralise the earth's field. A second weaker magnet may then be used to regulate the strength and direction of the field within the coil of the galvanometer. A galvanometer made sensitive in this way is very sensitive to slight magnetic disturbances either caused by actual changes of the terrestrial forces, or to disturbing currents (electric trams) or to the accidental displacements of magnetic material (keys, spectacles, corset steels) which are almost unavoidable in a laboratory. To get rid of that portion of these effects which is nearly uniform throughout the space occupied by the galvanometer needles, so called astatic magnetic systems are often used. These consist of two magnets or two sets of magnets with nearly equal magnetic moments, rigidly connected together, so that the similar poles of the two sets point in opposite directions. One of the sets is placed in the centre of the galvanometer coil, the other above or below it, and in some instruments the second set is placed in the centre of a second galvanometer coil in all respects similar to the first, but with the current passing round it in the opposite direction. The whole of the combination of magnets, called an astatic system, will behave like a very weak magnet towards the earth's force, and may set in a direction which it is impossible to predict.

Directing magnets have therefore to be used with an astatic system to bring it into its proper position with respect to the galvanometer coils and to regulate the strength of the field.

The increased sensitiveness of an astatic galvanometer depends on the fact that while the two sets of magnets oppose each other in so far as the earth's directing force is concerned, the electric current acts on both in the same direction. If there are two galvanometer coils, the current is led through them in opposite directions, so that the couples exerted by the currents in both coils have the same direction. A galvanometer with two coils can also be used as a "differential galvanometer," when it is required to test the equality of two currents or to measure very small differences between them. The two currents are in that case sent separately through the two coils, so that their effects on the suspended system of magnets oppose each other.

In thus increasing the sensitiveness of a galvanometer the time of oscillation of the needle is increased. This to a great extent counteracts the advantage gained by the increased sensitiveness, especially when owing to disturbing causes the zero of the needle is a little unsteady, for it becomes in that case impossible to distinguish a true deflection from an accidental shift. In modern galvanometers intended to be highly sensitive the suspended magnetic system is made as light and small as possible in order to secure a small moment of inertia and thus allow a corresponding decrease of the directing field without increase of the time of oscillation. The magnetic moments of the small magnets used are for the same reason made as large as possible by magnetising the steel to a high intensity*.

In the case of a d'Arsonval or suspended coil galvanometer the sensitiveness can be increased by increasing the number of turns in the coil or by increasing the strength of the field in which the coil is suspended, or by decreasing the torsional couple due to the supporting strip.

The time of oscillation is increased by the weakening of the couple due to the suspension, and must be kept down by a corresponding decrease in the moment of inertia of the coil.

Careful attention should be paid to the optical arrangement on which the measurement of the angles depends. If the objective method is used a good image of a wire on the scale should be secured in order to increase the accuracy of reading. Doubling the accuracy of reading doubles the effective sensitiveness of the galvanometer without increasing the periodic time. The image formed should be as perfect as the size of the mirror will allow; the latter should be either plane or concave with a radius of curvature of about one metre. The curvature of the mirror, and the definition of the image formed by it, may be investigated by using a source of light some distance away, and determining whether there is a good reflected image within a few feet of the mirror. If there is,

* See *Dict. App. Physics*, vol. II, "Galvanometers."

the distances of object and image will determine the radius of curvature. If the mirror is plane, or nearly so, the image may not be real or may be too far away, and in that case must be looked at through a telescope. The telescope being focussed for the image, we may easily determine the distance of a point which when looked at directly is also in focus. From this and the distance of the source of light the focal length of the mirror may be determined.

The principal adjustments of mirror galvanometers consist, according to the above explanations, (1) in securing that the suspended system can turn freely round a vertical axis through the largest angles on either side of the zero position which are likely to be used during the observations, (2) in securing that the resultant force on it with no current shall be nearly parallel to the plane of the galvanometer coils, and (3) in adjusting the strength of the field to the required intensity.

If it is desired to adjust a galvanometer *ab initio*, the first step should be to place the galvanometer on a firm support, if possible in the centre of the room so that access may be had to it from all sides. Freedom of motion of the galvanometer needle should be obtained by adjusting the level of the instrument. If it cannot be secured thus, carefully notice what is the cause of the impediment. Possibly the needle hangs too low or too high, and in that case the suspension must be altered, but this should only be done by experienced hands. When the needle is free, observe its approximate position of rest. With non-astatic systems it should be in the direction of the resultant field at the place of observation. If this is not the case, the fibre is probably twisted. Unless the angle of twist is very great, no serious error will result; but instruments should be provided with some means of turning the suspension head, and thus untwisting the fibre so that the needle may hang in the proper direction. It is always an advantage to be acquainted with the peculiarities of each instrument, and the effects of torsion should be noted by twisting the upper end of the suspension through a measured angle and noting the angle through which the suspended magnet turns in consequence. This angle of twist will prove useful in the future use of the instrument, because if the suspended system is losing its magnetisation, the angle will gradually increase and shew when the time has arrived for taking out the suspension and remagnetising the magnets.

If the instrument is of the suspended coil type it should be levelled so that the coil is midway between the poles, and the screw at the head of the suspension should be adjusted till the plane of the coil bisects the poles of the magnet.

For the suspended needle galvanometer, if M is the magnetic moment, I the moment of inertia, T the time of oscillation of the magnets, H the strength of the field, and N the torsion per unit angle of twist, we have the relation

$$T = 2\pi \sqrt{\frac{I}{HM + N}},$$

and if a twist through an angle α at the upper end of the suspension has produced an angular displacement θ of the magnet system:

$$N(\alpha - \theta) = HM \sin \theta,$$

where α will in general be sufficiently great compared to θ to allow the latter to be neglected on the left-hand side of the equation. If the suspension has been put together in the Laboratory, the weights and dimensions of different portions of it should have been noted, so that the moment of inertia can be calculated with sufficient accuracy. In that case, the above relations will allow the determination of M and N.

The value of N is of some interest in instruments intended for delicate work, for it determines the limit beyond which it is not possible to push the sensitiveness of the galvanometer, even if the earth's field be wholly neutralised.

In the case of the suspended coil instrument, if the damping is not very great, the time of oscillation may be obtained from the above equation by putting $M = 0$.

The time of oscillation and the direction of the needle having been noted, the position where it is desired to set up the galvanometer must be considered. This will generally be along a wall of the Laboratory, or at right angles to it, and therefore not necessarily in the direction of the magnetic meridian. A rough calculation will give the strength of the required magnetic field which together with that of the earth will set the magnet of the suspended type of instrument parallel to the galvanometer coils, and give a time of oscillation which should not be greater than 8 or 10 seconds. If the focal length of the mirror is about a metre, the scale may be set up at that distance from the mirror, but if the mirror is plane a convex lens of one metre focus should be fixed to the galvanometer as near to it as possible. If the mirror is neither plane nor has the required radius of curvature, the focal length of a lens must be calculated so that when fixed near the mirror, a point of light at a distance of one metre will give a real image at the same distance. The scale is now set at the proper distance, and the last adjustment of focus is made either by altering slightly this distance or by introducing a weak lens to alter the divergence of the beam incident on the mirror. A thin wire placed in the incident beam, moveable for greater facility of focussing, serves as the object, the image of which on the scale determines the angular position of the mirror.

Very often the preceding adjustments have been made for the student, at any rate approximately, and in that case his first care should be to check the adjustment for focus and to improve it when necessary. If the instrument is to be used not simply as an indicator but to measure deflections, the position of the scale must also be looked to. It should be placed at right angles to the line joining its centre to that of the mirror. This is most easily done, as explained in Section XXXI, by measuring the distance from the ends of the scale to the mirror, or if the galvanometer is covered by a glass shade, it will be sufficient to measure the distances from that point of the shade which is estimated to lie on the central line. The investigation on p. 156 shews that if an accuracy of one in 1000 in the readings is aimed at, the difference in the distances of the ends of a scale 50 cms. long from the mirror should not exceed 2 mms. for a mirror 100 cms. from the middle of the scale.

The last adjustments relate to the sensitiveness of the instrument. If this is insufficient in the case of a suspended magnet instrument a permanent magnet should be placed parallel to the galvanometer needle and so that the line joining the centres of the magnet and needle is approximately perpendicular to their magnetic axes. If the poles of the outside magnet point in the same direction as those of the suspended needle, the magnetic field will be weakened by it. The magnet is first placed at a distance and then brought slowly nearer, the spot of light being watched at the same time, so as to keep it near the centre of the scale. It will be found that when the outside magnet is brought too near, the galvanometer needle becomes unstable and is eventually reversed. When the magnet is placed so far away that the needle is just stable, the instrument is as sensitive as it can be made, but it will probably not be advantageous to use it in this most sensitive condition owing to the increase in the length of the time of oscillation.

In some instruments the controlling magnet is clamped to a vertical rod fixed to the galvanometer stand, and its position is altered by loosening the clamp and sliding the magnet upwards or downwards. Some makers have adopted a method in which two controlling magnets at a fixed distance are used. The field due to the two magnets may be strengthened or weakened by altering the angle between their magnetic axes. In the case of galvanometers having astatic systems, the sensitiveness may be altered in the same manner by an outside magnet.

Delicate galvanometers may be damaged when too strong a current is sent through them; and very often it is not possible at the beginning of an experiment to make sure that the electromotive force at the terminals of the galvanometer does not exceed the limit of safety. In that case a "shunt" should be used. This is a resistance smaller than that of the galvanometer and connected to it in such a way that the current will pass through the galvanometer and shunt in parallel. Sometimes a number of shunts are supplied with the instrument having their resistances graduated so that only the $\frac{1}{10}$, $\frac{1}{100}$ or $\frac{1}{1000}$th part of the current may pass through the galvanometer.

A shunt has been designed by Ayrton and Mather which to a close approximation will secure this for any galvanometer.

The sensitiveness of a galvanometer is measured by its power to indicate small variations of current, but it does not follow that the most sensitive galvanometer is the one that indicates the smallest variations of *the quantity to be measured.* If the annular space which is to contain the galvanometer coil is given, we may fill it either with thin or with thick wire; in the former case we shall get a "sensitive" galvanometer, but one having high resistance, while the thick wire will give a low resistance but smaller sensitiveness. It is on the relation between the resistance and the sensitiveness that the behaviour of the galvanometer depends. If we replace a thick wire by one of half the diameter, we replace each turn by four of the thinner wire, which have sixteen times the original resistance. Hence we have increased the sensitiveness four times and the resistance sixteen times. It may be proved quite generally that—neglecting the space lost by the insulation of the wire —the sensitiveness of galvanometers having the identical spaces filled by windings will vary as the square root of the resistance. In practice the insulation reduces the power from $\frac{1}{2}$ to $\frac{2}{5}$.

Writing κG^n for the sensitiveness of a galvanometer, where G is its resistance and κ a constant depending on the shape and size of the space containing the windings, $n = \frac{2}{5}$ to $\frac{1}{2}$ in a circuit containing an electromotive force E, the galvanometer and a resistance R, the current will be $E/(R + G)$ and the deflection of the galvanometer $\kappa EG^n/(R + G)$. The deflection of the galvanometer will therefore be very small when the resistance of the galvanometer is very small and also when it is very large, and it may be shewn that the deflection will be a maximum, when $G = Rn/(1 - n)$. Hence when small electromotive forces have to be measured, that galvanometer will be most suitable which has $\frac{2}{3}$ the resistance of the external resistance. For work with circuits which have small resistances, so called "sensitive" galvanometers will not be suitable, and low resistance galvanometers will give the best results. When the circuits are more complicated the above investigations have to be extended to find the most suitable galvanometer, but the general rule will be found to hold, that when the external resistances are small, the galvanometer resistance should be small.

In the case of galvanometers in which the poles of the suspended magnets come close up to the windings and in the case of suspended coil galvanometers, it cannot be assumed that the currents are proportional to the angles of deflections, even when these are small. If therefore small currents passing through the instrument are to be compared with each other, the instrument should be calibrated. This is done by the Potentiometer Method of Section LXIII.

A short account may in conclusion be given of the "damping" of galvanometers. A current passing through the instrument will deflect the needle, which will begin to oscillate about its new position of equilibrium. The oscillations follow the same laws as those of a pendulum, and will gradually diminish owing to frictional resistance and electrical damping.

In many cases it is desirable that the needle should come to rest quickly, for which purpose it is advisable to increase this resistance, either mechanically by attaching a vane to the suspended system, or electrically by bringing a mass of well-conducting material (copper) near the oscillating magnets. The currents induced in the copper by the motion of the magnets react on the latter and oppose its motion (Lenz' law).

When the damping is so great that the needle will not oscillate at all, but gradually takes up its new position of equilibrium, without passing beyond it, the motion of the needle is called "aperiodic," and the galvanometer is said to be "dead beat."

When the instrument is used to measure the quantity of electricity conveyed by a large current of short duration (discharge of a condenser, induction kicks), we require the deflection which would be produced if there were no friction, and in that case the damping should be small, and must be taken into account (see Section LXVII). A galvanometer used in this fashion is called a ballistic galvanometer.

The angular displacement of a dead beat galvanometer is represented by the expression

$$c\,(1 - e^{-\lambda t}),$$

where c is the ultimate displacement, which, as is seen, will be reached theoretically only after an infinite time.

It is generally found most convenient when a galvanometer is not used ballistically to adjust the damping till the motion is nearly dead beat. For this purpose the magnets of moving magnet galvanometers are sometimes provided with an aluminium vane moving in a closed box with little clearance. The air currents produced damp the motion, and the damping may be altered by moving the ends of the box towards or from the vane. In moving coil galvanometers the damping is diminished by putting additional resistance in circuit in series with the instrument, and increased by putting a shunt across its terminals.

The angular displacement from its position of equilibrium of a damped needle whose moment of inertia is I and magnetic moment M*, suspended by a fibre of torsional couple N, oscillating through a small angle in a uniform field H, and subject to a frictional couple whose moment is $2B$ times the angular velocity of the needle, is represented by the equation

$$x - c = ae^{-\frac{B}{I}t} . \sin\sqrt{\frac{MH + N}{I} - \frac{B^2}{I^2}}.t,$$

where t is measured from the instant when x has the value c.

If T is the time of oscillation, we may write

$$x - c = ae^{-\frac{B}{I}t} \sin\frac{2\pi t}{T},$$

where

$$\frac{2\pi}{T} = \sqrt{\frac{MH + N}{I} - \frac{B^2}{I^2}}.$$

* In the case of a moving coil galvanometer $M = 0$ and B consists of two parts, one due to air resistance, the other due to the action of currents induced in the circuit by the motion of the coil. See *Dict. App. Physics*, vol. II, "Galvanometers."

The maximum values of x will occur when $t = \frac{T}{2}\left(n - \frac{1}{2}\right)$ very nearly when n is an odd integer 1, 3, 5 ..., and the minimum values when $t = \frac{T}{2}\left(n - \frac{1}{2}\right)$ very nearly when n is an even integer 2, 4, 6 ....

Thus the ratio of a maximum elongation to the next in the same direction is equal to $e^{\frac{BT}{I}}$, and to the next in the opposite direction to $e^{\frac{B}{I}\cdot\frac{T}{2}}$.

Hence the difference of the logarithms to the base e of successive swings is constant. This constant is known as the "logarithmic decrement" of the oscillation and is written λ. Thus

$$\frac{B}{I}\frac{T}{2} = \lambda \text{ or } \frac{B}{I} = \frac{2\lambda}{T},$$

and the displacement x of the needle at time t may be written

$$x - c = ae^{-\frac{2\lambda t}{T}} \sin \frac{2\pi t}{T}.$$

When the first swing of a galvanometer needle is observed and it is desired to calculate what that swing would have been in the absence of all damping, we require to know the value of λ. As the time taken by the needle to pass from the position of rest to its first elongation is $T/4$, the above equations give $x_1 - c = ae^{-\frac{\lambda}{2}}$, where x_1 is the observed elongation and a the required amplitude. Hence

$$a = (x_1 - c)\, e^{\frac{\lambda}{2}} = (x_1 - c)\left(\frac{x_1 - c}{x_3 - c}\right)^{\frac{1}{4}},$$

where x_3 is the next turning point on the same side as x_1. If λ be smaller than ·1 this reduces to

$$a = (x_1 - c)\left(1 + \frac{\lambda}{2}\right) = (x_1 - c)\left(1 + \frac{1}{4}\frac{x_1 - x_3}{x_3 - c}\right),$$

an equation which we shall have occasion to use.

In case the numerical value of λ has to be determined to a higher degree of accuracy several successive turning points have to be observed, and from these the swings $x_0 - x_1$, $x_2 - x_1$ &c. calculated. We then have

$$\log_e (x_0 - x_1) - \log_e (x_n - x_{n+1}) \quad = n\lambda,$$
$$\log_e (x_2 - x_1) - \log_e (x_{n+2} - x_{n+1}) = n\lambda,$$
$$\&c., \&c.,$$

so that a series of independent value of λ may be obtained the mean of which is taken for the final result.

Exercise I. Set up a suspended magnet mirror galvanometer in the given position and measure the time of oscillation and the logarithmic decrement of the suspended magnet when oscillating under the action of the earth's force only. By means of an external magnet increase and decrease the time of oscillation to about 3, 5, 10, 15 and 20 seconds, and determine in each case the time, the logarithmic decrement, and the value of $2\lambda/T$.

Exercise II. Determine the deflection produced by a Daniell cell with a megohm in series with the galvanometer and calculate the current necessary to produce deflection of 1 mm. on a scale 1 metre away when the times of oscillation are 10 and 5 seconds, respectively.

The electromotive force of the Daniell cell may be taken as 1·07 volts.

Exercise III. Set up a moving coil galvanometer and determine its time of oscillation and logarithmic decrement on open circuit. The initial deflection may be produced by connecting the instrument through a key to a coil into which a magnet may be inserted. The key contact should be broken immediately the deflection has been produced.

Insert a resistance box in the circuit instead of the key, and determine the greatest value of the resistance for which the movement of the coil is dead beat.

Insert double this resistance, place a second resistance as a shunt across the galvanometer terminals and determine the value of the second resistance for dead beat motion. Make the series resistance four times its original value and again determine the shunt resistance which gives dead beat motion. From the three observations shew that for dead beat motion the joint resistance of the two circuits in parallel across the terminals of the galvanometer is constant.

Exercise IV. Determine, as in Exercise II, the current which would produce a deflection of 1 mm. on a scale 1 metre from the instrument.

SECTION LIII

THE POST OFFICE RESISTANCE BRIDGE

Apparatus required: *Galvanometer and scale, Post Office resistance box, Leclanché cell, coils, voltmeters, platinoid wire, and connecting wires.*

The student is supposed to be familiar with the principle of a resistance bridge, and to have had some practice in the measurement of resistance by the simple form of bridge.

When two of the resistances of the bridge are formed by a stretched wire, with a moveable sliding contact for the junction leading to the battery or galvanometer as in the simple form of bridge, the accuracy attainable is of the order 1 or 2 per cent. When greater accuracy is required a bridge consisting of a number of resistances, connected in series, and compactly placed together in a box which is known as a "Post Office box" (Fig. 107 *a* and *b*) is used.

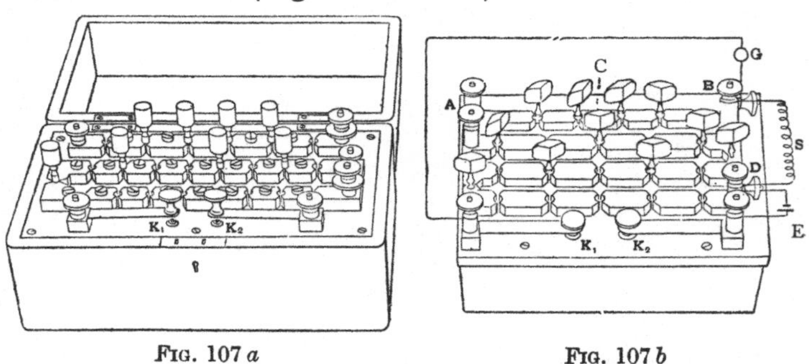

Fig. 107 *a* Fig. 107 *b*

The connections to the resistances are shewn in the diagram of the bridge (Fig. 108). The battery and galvanometer circuits are brought to spring contacts at K_1, K_2, placed on the box as shewn in Figs. 107. There is a metallic connection inside the box between C and K_2 and between A and K_1 as shewn in Fig. 108. The letters in Figs. 107 *b* and 108 correspond. The resistance to be measured is placed between B and D, and the battery and galvanometer connections are made as shewn in the figure. The arms AC and BC each consist in general of resistances of 1000, 100 and 10 ohms, so that P and Q (Fig. 108) may either be equal in three different ways; or in the ratio of 10 : 1 in two different ways; or in the ratio of 100 : 1 by making one resistance equal to 1000 and the other equal to 10 ohms. According to the resistances placed in the arms P and Q, to balance the bridge the resistance in the arm R should be S, $10S$ or $100S$, if P is equal to or greater than Q, and $\cdot 1S$ or $\cdot 01S$ if P is smaller than Q. Another variation can be made by interchanging the battery and galvanometer, and it is therefore seen that a great many combinations for producing balance in the bridge are available. But these combinations are not equally good, the unknown resistance being capable of measurement with greatest accuracy for a particular choice of P and Q, depending on the value of the resistance to be measured and on the resistances of the galvanometer and battery circuits. In nearly all cases in which the highest obtainable accuracy is not required the following rules, if attended to, will prove a sufficient guide to the student in the selection of a proper combination of resistances*.

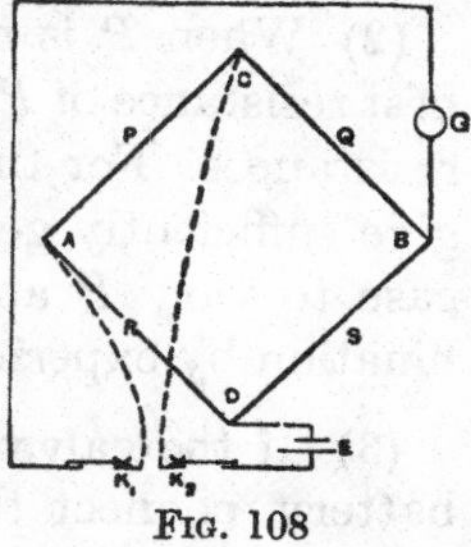

FIG. 108

(1) As the resistance R between A and D in the ordinary P.O. box can be varied in steps of one ohm, and as one-tenth of that amount may with certainty be estimated by interpolation of galvanometer deflections, resistances of over 100 ohms can be measured with an accuracy of at least $\cdot 1$ %, if the arms P and Q are made equal to each other. Hence if a rough measurement has shewn that S has a value of more than 100 ohms, make P and Q equal to each other. If S lies between 100 ohms and 10 ohms and has to be measured to less than 1 %, P should be 10 times as great as Q, while if

* See Smith, *Dict. App. Physics*, vol. II, "Resistance Standards."

S has a smaller value than 10, P should be 1000 ohms and Q 10 ohms.

(2) When P is equal to, or ten times as great as Q, the best resistance of P depends on the galvanometer and battery resistances. For the exercises in this section, any value will give sufficiently good results. It is easier in each particular case to vary P and Q and find the most favourable combination by experiment, than to determine it by calculation.

(3) If the galvanometer has a greater resistance than the battery, connect the galvanometer circuit so as to join the junction of the two greatest to that of the two least resistances. If the battery has a greater resistance than the galvanometer, the battery ought to be placed between the junction of the two greatest and that of the two least resistances.

Thus if the galvanometer has the greater resistance, and if $P = 100$, $Q = 10$ and $S = 8$ so that $R = 80$, the galvanometer circuit should be connected to B and K_1, and the battery to D and K_2, as shewn in Fig. 108. Whenever P and Q are equal, the best connection is that in which the battery and galvanometer circuits are different to that shewn in Fig. 108.

Exercise I. Measure to the nearest ohm the given resistances. The object of this exercise is to make the student familiar with the connections of the P.O. box, and to provide practice in determining a resistance quickly when the highest accuracy is not required.

If the image of the wire on the scale is not sharp make the necessary adjustments of lamp and scale. Make connections to the galvanometer and Leclanché cell as shewn in Fig. 108. Rule columns in your note-book and record the zero reading of the galvanometer, as indicated in the table below. Make P and Q each equal to 1000 ohms. Take a resistance of 1000 ohms out of the bridge arm AD, and to begin with, leave the resistance to be measured disconnected.

Press the key K_2 to make the battery circuit, then for an instant the key K_1 to make the galvanometer circuit, and release first K_1, then K_2. Observe in which direction the spot of light moves, and note that in the subsequent measurement a deflection in the same direction will always mean that S is larger than R; *i.e.* that the resistance in the arm AD is too small. If the motion is irregular, adjust the galvanometer according to the instructions in Section LII.

Now connect the resistance to be measured as shewn in the figure, and see that all plugs which are not taken out of the box are firmly in their places, and that all screw contacts are clean and secure. Again press down K_2 and then K_1, observing in which direction the galvanometer begins to move. Release K_1 quickly, so as to avoid passing a possibly large current through the galvanometer for a longer time than is necessary.

If the galvanometer is deflected in the same direction as before, this shews that the resistance to be measured is greater than 1000, because R being 1000, the deflection is in the same direction whether S is equal to the given resistance or infinitely large. In that case a higher value should be tried in the arm AD, say 5000, and so on until a resistance has been found which gives a deflection in the opposite direction. Having thus ascertained by trial that the resistance lies between say 1000 and 5000 some intermediate resistance should be tried, say 3000, and the direction of deflection will then shew between which limits the right value for balance lies. A succession of trials each time approximately halving the interval will quickly reduce the limits, until two values of R are found differing from each other by one ohm. The one which gives the smallest deflection will give to the nearest ohm the correct resistance of S. Similarly should S have been found to be smaller than 1000, successive trials of 500, 250 &c. will ultimately give a lower limit, and when this is found a successive halving of the interval will again give the required resistance.

The chief precaution to be taken by students is to avoid confusion as to the meaning of the deflections to one side or the other. As soon as it has been ascertained to which side the deflection takes place when the resistance taken out is too small, i.e. the direction of motion of the spot of light when S is not inserted in the circuit, this direction must be carefully and conspicuously noted so that no mistake will afterwards be made.

The reason for pressing down the keys K_2 and K_1 in succession is to avoid the effects of the self-induction of the resistances or galvanometer, which may cause a different distribution of currents at the instant of making the battery circuit to that which obtains afterwards. Pressing the key K_2 when the connections are as in Fig. 108 completes the battery circuit, and as the period during which the current is affected by self-induction is very short, K_1 can be pressed down almost immediately.

Before an approximate balance has been obtained, the galvanometer key should only be pressed down for a sufficient time to shew in which direction the needle moves, as it is desirable to avoid as much as possible the passing of unnecessarily large currents through the galvanometer. When the balance is nearly right the key is kept down until the deflection can be read off either by noting the amplitude of the first swing or by waiting till the needle has come to rest.

Determine the resistance of the coils provided and arrange your results as follows:

Date:

Galvanometer No. 2 (Resistance = 67 ohms), Resistance box C.

$P = 1000. \quad Q = 1000.$

R	Zero	Deflected reading	Deflection
1000 ($S = \infty$)	50·1	to right	+
1000		to left	–
500		"	–
200		to right	+
400		51·4	+ 1·3
450		46·2	– 3·9
420		49·2	– ·9
410	50·1	50·3	+ ·2
412		50·1	zero
413		50·0	– ·1
411	50·1	50·2	+ ·1
$S = R = 412$ ohms.			

Until a student has obtained sufficient practice, his Book of Observations should contain a statement of all his observations as above. In writing out his results one example should be given in full; for the other resistances measured, the results only need be given.

Exercise II. Determine the same resistances to the greatest accuracy which the apparatus at your disposal will allow.

The resistances having been approximately determined, the student should for each of them separately consider the rules mentioned at the beginning of this section and fix on the values of P and Q which he considers most suitable. The rule about the proper connection of galvanometer and battery should also be attended to. When the galvanometer and battery circuits are interchanged, the order in which the keys K_2 and K_1 are pressed down must of course be reversed also. From his previous results he will at once be able to take out of the arm AD the plugs necessary to adjust the balance almost correctly. A few further trials, if necessary, will then again lead him to the two values R and $R + 1$ between which the balance lies. The deflections for both these must then be carefully observed on the scale, and by interpolation one, and possibly two decimals may be found.

If the deflections are not sufficiently large, the galvanometer must be made more sensitive. Resistances of not less than five ohms should in each case be accurately determined to ·1 % and if possible more accurately. To ensure this accuracy, however, with the smaller resistances it will be necessary to attend carefully to the connections and to make sure that the plugs in the resistance box are firmly in their sockets.

If the resistances are of copper or some other unalloyed metal, the resistance of which increases about 1 % for every 3° C., the temperature should be noted, and the current should be passed through the coil for as short a time as possible so as to avoid heating effects, otherwise the resistance will be found to increase gradually. To make sure that no such change is taking place, the observations should be repeated. Owing to the temperature effect, the accuracy obtainable with pure metals is much smaller than with alloys, which have a small temperature coefficient.

Enter your results as follows:

	Material	Q	P	Junction between P and Q connected to	R	Temp. of coil
Coil No. 1	Copper	1000	1000	Galvanometer	412·3	16°·2
,, ,, 2	Platinoid	100	10	Battery	68·25	16°·4
·	·	·	·	·	·	·
·	·	·	·	·	·	·
,, ,, 5	Manganin	100	10	Battery	7·423	15°·9

Exercise III. To determine the resistivity of the material of a wire.

Measure the resistance of the wire provided to ·1 %, noting the length of the wire under the screws which clamp it to the resistance box. Measure the diameter of the wire in four places and the total length. Subtract the length under the clamping screws from the total length, and calculate the resistivity.

Arrange your results as follows:

Date:

Resistivity of Platinoid.

Total length of wire ... ... = 100·2 cms.
Length under screws... ... = 1·7 ,,
Length (l) used ... ... = 98·5 ,,
Diameters at different places by screw gauge = ·0342, ·0344, ·0346, ·0348.
Mean diameter ... ... = ·0345 cm.
Mean cross section (a) ... = ·000937 sq. cm.
Resistance (R) ... ... = 4·205 ohms.
Resistivity $= \frac{Ra}{l}$ ... ... $= 40{\cdot}0 \times 10^{-6}$ ohms per cm. cube at 18° C.

SECTION LIV

HIGH RESISTANCES

Apparatus required: *High resistances, one about* 30,000 *ohms, another about a megohm, insulating material, resistance boxes, cells and high resistance galvanometer.*

When the resistance to be measured is large compared to that of the most sensitive galvanometer available, the resistance bridge loses its advantages and ceases to give more accurate results than simpler and more direct methods. When the accuracy required does not exceed about ·1 % these simpler methods will be sufficient.

If a standard resistance of approximately the same value as the one to be measured is available, the method known as the "direct deflection method" may be used.

Let a cell of resistance B ohms be connected in series with a high resistance of R_1 ohms, and a mirror galvanometer of resistance G ohms (Fig. 109).

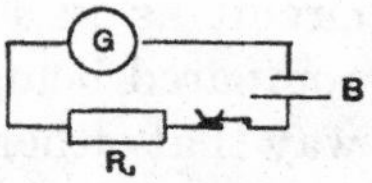

FIG. 109

The current C_1 through the circuit is given by

$$C_1 = \frac{E}{B + G + R_1}.$$

Similarly when a second known resistance R_2 is substituted for R_1,

$$C_2 = \frac{E}{B + G + R_2},$$

so that

$$B + G + R_2 = \frac{C_1}{C_2}(B + G + R_1).$$

The battery resistance will in nearly all cases be quite negligible compared to the high resistance to be measured, and the resistance of the galvanometer will be known or must be determined by an independent measurement. Hence R_1 being known, R_2 can be calculated from the above equation, if the galvanometer deflections d_1 and d_2 corresponding to C_1 and C_2 are observed, and the ratio of the currents deduced from the ratio of the deflections. If the indications of the mirror galvanometer have been calibrated (p. 271) a table is probably available in the Laboratory by means of which deflections may be converted into currents. In the following exercise it is assumed that results of sufficient accuracy may be obtained by taking the deflections to be proportional to the currents. In that case the unknown resistance R_2 is given by

$$R_2 = (R_1 + G)\frac{d_1}{d_2} - G = R_1\frac{d_1}{d_2} + G\frac{d_1 - d_2}{d_2},$$

and if G is small compared to R_1 and R_2

$$R_2 = R_1\frac{d_1}{d_2}.$$

Taking a resistance box having a total resistance of not less than 10,000 ohms as R_1, determine the resistance R_2 of the coil provided. Arrange the circuit as in Fig. 110, the resistances to be compared being placed side by side in such a way that either one or the other may be put in circuit with the battery and galvanometer.

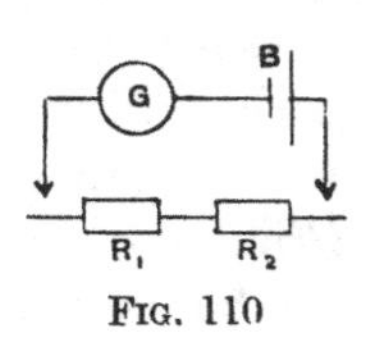

FIG. 110

If the deflections of the galvanometer with R_1 in circuit are too large to enable observations to be taken, diminish the sensitiveness of the instrument by an external magnet till they are within the limits of the scale. If this is not possible arrange the circuit as in the second method described below.

If the resistance to be determined is more than three or four times that of the standard, the ratio of the currents will not be capable of measurement to a sufficient degree of accuracy by the direct deflections. The method should then be modified by the addition of an arrangement which allows the electromotive force to be varied in such a way that the deflections d_1 and d_2 are not very unequal.

For this purpose a cell of small resistance B is connected to the ends of a resistance box containing r (about 1000) ohms. The galvanometer in series with the high resistance to be measured is connected across part only (r_1) of the resistance (Fig. 111), and this part may be varied. A resistance box provided with the necessary terminals and the plugs for varying r_1 is known as a Volt Ratio Box. Such boxes may have $r = 10{,}000$ or $100{,}000$ ohms.

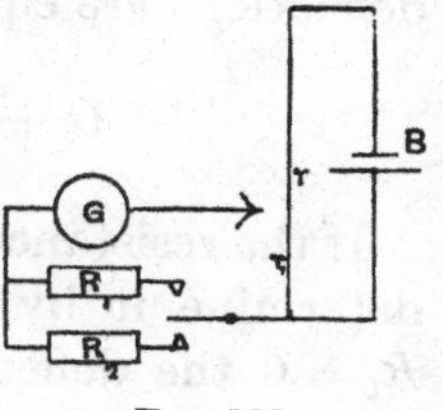

Fig. 111

Let the resistance to be measured be R_2, the standard resistance R_1, and the galvanometer resistance G.

Let $B + r = r'$ be the total resistance in the cell circuit, and r_1 that part of it between the terminals of the galvanometer circuit when R_1, and r_2 when R_2, is tested, the currents through the galvanometer being in the two cases C_1 and C_2. Then if E is the electromotive force of the cell a simple application of the laws of derived circuits gives

$$\frac{G + R_1 + r_1 - \dfrac{r_1^2}{r'}}{\dfrac{r_1}{C_1}} = \frac{G + R_2 + r_2 - \dfrac{r_2^2}{r'}}{\dfrac{r_2}{C_2}} = \frac{E}{r'},$$

from which we find

$$G + R_2 = \frac{C_1}{C_2} \cdot \frac{r_2}{r_1}\left(G + R_1 + r_1 - \frac{r_1^2}{r'}\right) - \left(r_2 - \frac{r_2^2}{r'}\right),$$

in which in general r may be substituted for r' without appreciable error.

Hence R_2 may be determined if G is known and we can find the ratio $\dfrac{C_1}{C_2}$ from the galvanometer deflections. If the currents are not proportional to the deflections the galvanometer must be calibrated. If the currents C_1 and C_2 produce deflections which when corrected for calibration errors are equal to d_1 and d_2, $C_1/C_2 = d_1/d_2$, and the above equation becomes

$$G + R_2 = \frac{d_1}{d_2} \cdot \frac{r_2}{r_1}\left(G + R_1 + r_1 - \frac{r_1^2}{r'}\right) - \left(r_2 - \frac{r_2^2}{r'}\right).$$

The calibration can be avoided and the arithmetical work shortened if the resistances r_1, r_2 can be so adjusted that the deflections are equal. In that case

$$G + R_2 = \frac{r_2}{r_1}(G + R_1) + \frac{r_2(r_2 - r_1)}{r'}.$$

If the resistance G of the galvanometer is unknown we may determine it by reducing r_1 to so small a value that with $R_1 = 0$ the deflection of the galvanometer is not too large to be read on the scale. After reading the deflection and correcting it for calibration error insert in series with the galvanometer a resistance which reduces the deflection, again corrected for calibration error, to half its former value. The resistance of the galvanometer is equal to the resistance added.

If the resistance G cannot conveniently be determined in this way we may eliminate it from our equations by taking a third observation with the known resistance changed considerably in value. If R_1' is its new value, r_1' that of r_1 and C_1' that of C_1, we have

$$\frac{G + R_1' + r_1' - \dfrac{r_1'^2}{r'}}{\dfrac{r_1'}{C_1'}} = \frac{E}{r_1}.$$

Hence

$$\frac{R_2 - R_1 + (r_2 - r_1)\left(1 - \dfrac{r_2 + r_1}{r'}\right)}{\dfrac{r_2}{C_2} - \dfrac{r_1}{C_1}} = \frac{R_1' - R_1 + (r_1' - r_1)\left(1 - \dfrac{r_1' + r_1}{r'}\right)}{\dfrac{r_1'}{C_1'} - \dfrac{r_1}{C_1}},$$

from which R_2 may be calculated.

If G is to be determined the most suitable values of R_1' and r_1 to use are discussed on p. 351.

Determine by this method the resistances of the blacklead line, and of the samples of ivory and red fibre provided.

From the dimensions of the samples calculate the resistivities of the materials.

For materials of higher resistivity it may be necessary to raise the electromotive force applied to the outer terminals of the Volt Ratio Box. The Box generally has engraved on it the highest electromotive force which may be safely applied to it.

Assuming the deflections to be proportional to the currents, tabulate your observations and results as follows:

Direct Deflection Method.

Date:

Galvanometer No. 2 (Resistance 6030 ohms).

Standard Resistance (R_1) = 10,000 ohms.

Resistance	Deflection
10000	508·7
R_2	274·5
10000	510·3
	$d_1 = 509{\cdot}5$
	$d_2 = 274{\cdot}5$

$$R_2 = 16430 \times \frac{509{\cdot}5}{274{\cdot}5} - 6030 = 24{,}460 \text{ ohms.}$$

Volt Ratio Box Method.

Galvanometer No. 2.

Standard Resistances from box C.

R	$\dfrac{r = 11{,}110 \text{ ohms}}{r_1}$	Deflection
5000 ohms	14 ohms	31·1
Pencil Line	1110 ,,	30·7
11,110 ohms	22 ,,	31·2

$$\therefore \frac{R_2 - 5000 + 1096\left(1 - \frac{1124}{11110}\right)}{\frac{1110}{30\cdot7} = \frac{14}{31\cdot1}} = \frac{6000 + 8\left(1 - \frac{36}{11110}\right)}{\frac{22}{31\cdot2} - \frac{14}{31\cdot1}},$$

$$\frac{R_2 - 3917}{37\cdot74} = \frac{6008}{\cdot255},$$

$$\therefore R_2 = 893{,}000 \text{ ohms.}$$

Similarly for the samples of insulating material, which may require the ratio r_2/r_1 to be made greater than was necessary in the above case.

SECTION LV

LOW RESISTANCES

Apparatus required: *Low resistances and standard low resistance, storage cell, resistance coil, two four-way keys, Daniell cell, resistance box and mirror galvanometer.*

When a resistance which is only a small fraction of an ohm is to be measured, the resistance bridge is no longer capable of giving accurate results owing to the uncertain resistances introduced at the various contacts being comparable with the resistance to be measured. Several special methods of measurement have therefore been devised, one of which is known as the "fall of potential method." It depends on the comparison of the differences of potential produced by the same current at the ends of the unknown resistance and of a known standard resistance.

The two resistances R_1 and R_2 (Fig. 112) are connected in

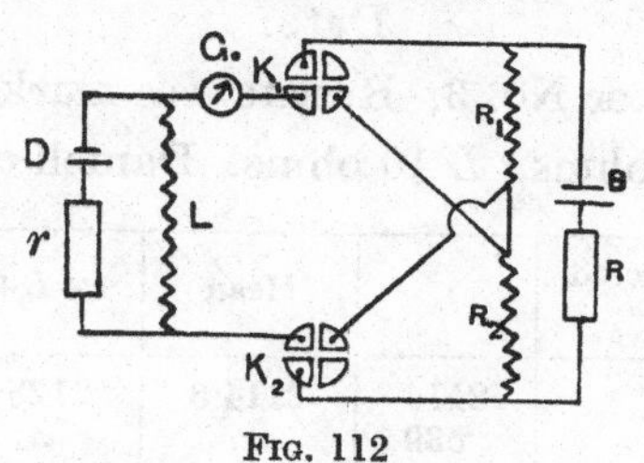

Fig. 112

series with a third resistance R of a few ohms, introduced to regulate the current, and to a storage cell B. From the ends of the two resistances R_1 and R_2, wires are taken to keys which enable each resistance to be connected through a galvanometer G in parallel with a known resistance L of about 10 ohms, through which and a resistance box r a Daniell cell D sends an independent current.

The resistance r is adjusted so that when L is connected to

R_1 there is no deflection of the galvanometer and its value noted. L is then connected to R_2 and r adjusted till there is no deflection, then again to R_1. If r_1 is the mean value for balance with R_1 and r_2 with R and r_0 is the resistance of the cell,

$$\frac{R_1}{R_2} = \frac{r_0 + r_2 + L}{r_0 + r_1 + L}.$$

If r_0 can be neglected in comparison with r_2 and r_1,

$$\frac{R_1}{R_2} = \frac{r_2 + L}{r_1 + L}.$$

Determine by this method the resistance of the three consecutive lengths marked A, B and C of the given wire, and compare the sum of the result with that obtained by testing the resistance of A, B and C in series.

The connections to the galvanometer may be arranged by means of two four-way keys, K_1 and K_2 as indicated in Fig. 112, the plugs being inserted to connect the galvanometer to the two resistances in turn. The resistance R should be sufficient to prevent r falling below 400 or 500.

Enter your observation as follows:

Date:

Galvanometer No. 3. Resistances marked A, B, C.

R about 5 ohms. L 10 ohms. Daniell cell 4 ohms.

Galv. connected up to	r	Mean	$r+L+r_0$	Ratio
$R_1 = \cdot01$	2114	2114·5	2128·5	
R_2 (A)	539	...	...	3·87
$R_1 = \cdot01$ ohm	2115	539	553	

Hence $A = \cdot0387$ ohm.

Similarly for the other given resistances, B and C.

Check the results by measuring the resistance $A + B + C$.

SECTION LVI

THE RESISTANCE OF A GALVANOMETER

KELVIN'S METHOD

Apparatus required: *Galvanometer, Post Office resistance box, Leclanché cell, resistance box of* 1000 *ohms and connecting wires.*

If the four arms of the resistance bridge (Fig. 108, p. 277) satisfy the relation $P/Q = R/S$ there is no difference of potential between the points A and B, and if they are directly connected by a wire, no current will pass through it. Hence the introduction of such a wire would not alter the strength of the current in any of the arms of the bridge. Conversely, we may conclude that if the current in any one of the branches is the same whether A and B are directly connected together or not the above relation is satisfied. If the branch R contains a galvanometer, the deflection of which serves to indicate the strength of the current through it, we may judge whether the resistances are balanced or not by making and breaking the contact of a wire connecting A and B. If the deflection is the same in both cases the resistance in the branch R must be equal to PS/Q and may therefore be determined without the assistance of a second galvanometer between A and B.

The above considerations lead to an interesting and useful method of determining the resistance of a galvanometer. To obtain accurate results it is in general necessary to send through the instrument a current which under ordinary circumstances would drive the spot of light off the scale. Reducing the sensitiveness of the galvanometer so as to make the deflection measurable would not get over the difficulty, because the test itself would become less sensitive. It is possible however to work with a large deflection and yet have the spot of light on the scale, because the zero reading of the galvanometer is not required and may therefore be outside the limits of the scale. In the case of a galvanometer of the suspended needle type, it is convenient to begin by deflecting the needle till the spot of light is at the end of the scale by means of a weak magnet arranged so as to alter the direction of the field at the needle without increasing its strength. The battery is then connected to a resistance r (Fig. 113, p. 293) so that only a portion of the current passes through the bridge, the spot of light is brought on to the scale by adjustment of the resistances r and r_1. r_1 is then increased, and the magnet moved so that it still brings the spot of light on to the scale without materially increasing the strength of the magnetic field at the galvanometer needle.

In the case of a d'Arsonval galvanometer, it is only necessary to rotate the whole instrument through an angle of about 30° and to adjust r and r_1, until the spot of light appears on the scale.

When the resistance Q of the bridge may be made smaller than the resistance of the galvanometer, the best arrangement is to have Q as small as possible, and P as large as it can be without R being caused to exceed the maximum resistance available in the arm AD. The battery should come between the points A and B, *i.e.*, the battery and short circuiting wire in the figure should be interchanged. When the lowest available resistance of Q is larger than the galvanometer resistance, Q and P should be made equal to each other and as large as possible, and the battery should then be connected as shewn in the figure. If the resistance of the galvanometer is not even approximately known the last-mentioned arrangement is the most suitable and the sensitiveness will generally be sufficient.

In the following exercise it is required to measure the resistance of a d'Arsonval galvanometer, the method for other galvanometers only differing as explained above in the means adopted to alter the zero reading.

Make connections as shewn in the figure. With r_1 small and r large, pass a current of short duration through the galvanometer by pressing down the key K_2 for an instant.

Turn the galvanometer through an angle of about 30° or 40° in the *opposite* direction to that in which the spot of light moved on making contact.

With $P = Q = R = 1000$ close the battery circuit at the key K_2 and adjust r_1 and r until the spot of light is on the scale. This is done most rapidly by watching the galvanometer mirror itself instead of the spot of light. Press the short circuiting key K_1 and observe the side to which the spot of light moves. Then make $R = 5000$ and repeat the observation. If the motion is in the same direction as the first it indicates that S lies below 1000 or above 5000. The first supposition being the more probable in the present case, repeat the observation with $R = 500$.

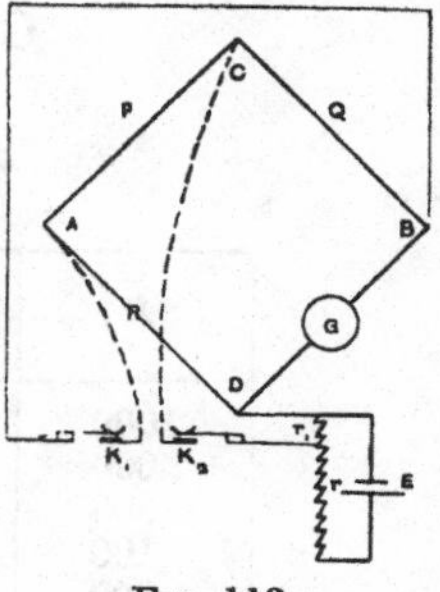

FIG. 113

Proceed as in Section LIII to find the value of R for which the spot of light does not move when the key K_1 is closed.

If no value of R satisfies this condition, find the values differing by unity for which the deflections are in opposite directions. Determine these deflections and find the value of R for accurate balance by interpolation.

If the resistance of the galvanometer is greater than the lowest available resistance of Q, which will be 10 with the resistance boxes in common use, connect the short circuiting wire to C and D through K_2, and the battery circuit to A and B through K_1. Make $Q = 10$ and $P = 1000$ if a resistance of $100\,S$ can be placed in the arm AD, otherwise make $P = 100$. Take out of the branch AD the resistance which according to the previous determination should produce balance; and if there is a deflection proceed to improve your result. When the balance is nearly perfect stop the current and wait a few minutes in order to allow the galvanometer coil to take up the temperature of the room, it having probably been heated by the passage of currents sufficiently to affect its resistance sensibly. Then repeat the observations.

Enter your results as follows:

Date:

Galvanometer d'Arsonval C.

Battery joining C and D.

$P = 1000, \quad Q = 1000.$

R	Reading with AB open	Reading with AB closed	Deflection
1000	15·1	+	large to right
5000	12·1	+	,, ,,
500	23·2	+	,, ,,
100	25·4	−	,, to left
300	24·0	41·3	+ 17·3
200	24·0	22·5	− 1·5
220	23·6	24·5	+ ·9
210	23·4	22·4	− 1·0

Resistance of galvanometer = 215 app.
Temperature of room = 15°·4 C.

The battery and short circuit were now interchanged and with $Q = 10$, $P = 100$ it was found that $R = 2154$ gave a balance but that this resistance seemed slowly to increase owing no doubt to the heating of the current. The same result was obtained after waiting five minutes. Hence for the final result:

Resistance of galvanometer C = 215·4 ohms.
Temperature = 15°·4 C.

SECTION LVII

THE RESISTANCE OF A CELL

LODGE'S MODIFICATION OF MANCE'S METHOD

Apparatus required: *Daniell and Leclanché cells, Post Office resistance box, high resistance galvanometer.*

If in the resistance bridge arrangements (Fig. 108, p. 277) the relation $PS = QR$ is satisfied, the arms of the bridge CD and AB are said to be conjugate to each other. No electromotive force in one of these arms will produce a current in the other, and no change of resistance in one of them will modify any current in the other produced by electromotive forces in one or more of the branches of the bridge.

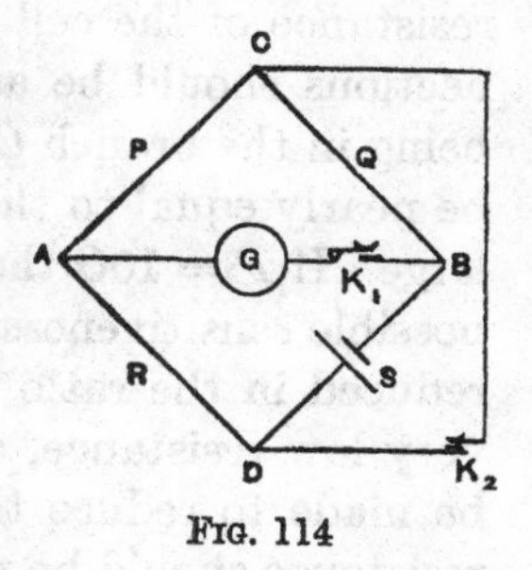

FIG. 114

This fact was first used by Mance to determine the resistance of a cell. He placed the cell in one of the arms, and replaced the battery, which in the ordinary resistance bridge supplies the current, by a simple key K_2 (Fig. 114). Owing to the presence of the cell in BD a current passes through the galvanometer, but when $P/Q = R/S$ this current should be the same whether the contact at K_2 is open or closed.

To prevent a large current passing through the galvanometer it should in the first instance be shunted by a small resistance which can be increased as the balance is made more accurately. The key K_2 should only be pressed down for one-quarter of the period of the galvanometer to allow the first deflection to be read. The closing of the circuit at K_2, though it does not affect the current in the galvanometer branch, will increase the current through the cell itself, and this increase is accompanied by changes in the resistance and electromotive force of the cell. This is a disturbing cause which affects all measurements of the resistance of a cell, but the difficulty may to some extent be overcome by allowing the key K_2 to be closed only for a very short interval of time.

The modification of Mance's method introduced by Lodge, gets over the difficulty by the introduction of a condenser in series with the galvanometer (Fig. 115). The condenser will be charged to a difference of potential equal to that between the points A and B; and whenever a sudden change in that difference of potential occurs, an instantaneous current will pass through the condenser. When however the relation $P/Q = R/S$ is satisfied, the galvanometer will not be affected by the sudden pressing down of the key K_2.

The most sensitive arrangement for the determination of the resistance of a cell is that in which Q is equal to the resistance of the cell and P is as large as possible. The connections should be as in Fig. 115, the short circuiting key being in the branch CD. It is more important that Q should be nearly equal to the cell resistance, than that P should be large. If $P = 10Q$ the arrangement has 90 % of the greatest possible sensitiveness, but when $Q = 10S$ the sensitiveness is reduced in the ratio of 10 to 1. Hence when the cell has a very low resistance, either some special arrangement should be made to reduce the value of Q, or an accurately known resistance should be placed in series with the cell, so that the total resistance of S is not less than half that of the lowest available value of Q.

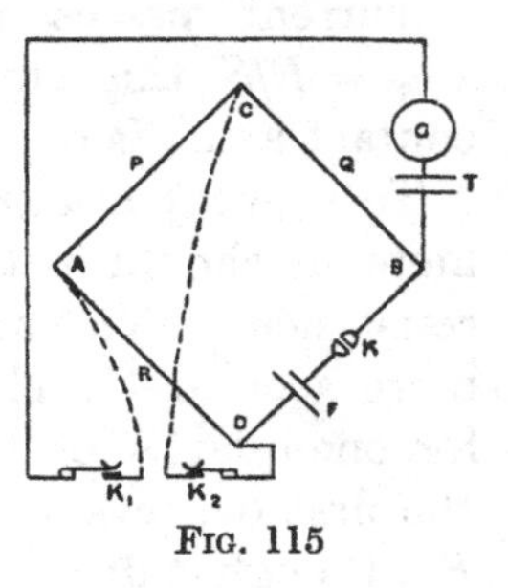

Fig. 115

Set up a Daniell cell, taking care that the zinc is not covered with a deposit of copper. Place it in the arm BD of the Resistance-Bridge Box $ABCD$ (Fig. 115), putting a plug-key K in circuit, in order that the current can be stopped except when observations are being taken.

Place a condenser T of about ·3 microfarad capacity in series with a high resistance galvanometer, and connect to the bridge as shewn. Make $Q = 10$, $P = 1000$, and choose for first trial a resistance R which you expect to be greater than that of the cell to be tested, say 100 in the present instance. Make circuit at K_1, then momentarily at K_2. Note the direction in which the spot of light moves. Reduce the value of R and proceed as in Section LIII to determine its value for balance. If the key K_2 is held down, the spot of light will be seen to drift owing to changes in the cell brought about by the passage of the current. Wait five minutes without passing a current through the cell, and determine the resistance again. Repeat again after a further five minutes.

Determine the resistance of the Leclanché cell in the same way, then placing the two cells first in series, then in opposition, determine their joint resistance.

Record as follows:

Date:

$P = 1000$, $Q = 10$ ohms.

1. Daniell cell, No. 3.
   Balance obtained with $R = 515$, *i.e.* $S = 5{\cdot}15$ ohms.
   After five minutes ... ... $= 5{\cdot}10$,,
   After further five minutes ... ... $= 5{\cdot}07$,,
2. Leclanché cell, No. 4.
   Approximate balance obtained for
   R between 470 and 480 ... $S = 4{\cdot}75$ approx.
   After five and ten minutes the resistance was approximately the same.
3. Leclanché and Daniell cells.
   In series ... ... $S = 9{\cdot}9$ ohms.
   In opposition ... ... $= 9{\cdot}8$,,
   Calculated resistance ... ... $= 9{\cdot}8$,,
4. Leclanché cell, No. 4 ... ... $= 4{\cdot}8$,,
5. Daniel ,, ,, 3 ... ... $= 5{\cdot}02$,,

In making the calculations under 3 use the means of 1 and 5, and of 2 and 4.

SECTION LVIII

COMPARISON OF RESISTANCE STANDARDS

Apparatus required: *Board with copper block mercury cups, two nearly equal resistances, standards to be compared, resistance box, mirror galvanometer, voltaic cell.*

When nearly equal resistances, as for instance a number of ohm standards, are to be compared together with great accuracy, a modification of the bridge method may be used.

The two "proportional arms" P, Q (Fig. 116) of the bridge are nearly equal resistances, P being less than Q by one part in 200 or 1000* according to the accuracy of the resistances to be compared, and the two standards R, S to be compared form the other arms. A resistance box r is connected in parallel with Q. The standards R, S are connected to the bridge by mercury cups in copper blocks in such a way that they can be readily interchanged.

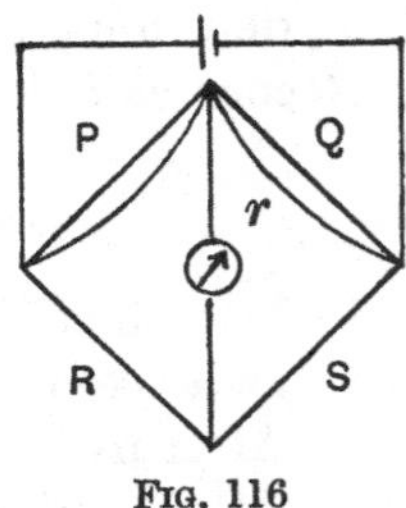

Fig. 116

Let a balance be found when R and S are placed as shewn in the figure, and r_1 is the resistance of the box in parallel with Q. Interchange R and S and let r_2 be the resistance of the box to produce a balance.

If r_1 or r_2 are comparatively small it may not be possible to get an exact balance with the resistances available in the box. In this case find two resistances differing by an ohm for one of which the deflection of the galvanometer is in one direction and for the other in the other direction. Observe the deflection and find by proportional parts the value of the resistance which would give a perfect balance.

* If P and Q are equal the necessary reduction of the resistance of P to allow all the adjustments for balance to be made at Q can be secured by a shunt of 200 to 1000 ohms connected permanently to P.

In all cases the balance should be found with the main current in each direction in turn through the bridge.

With the resistances R and S, as in Fig. 116, we have

$$\frac{R}{S} = P\left(\frac{1}{Q} + \frac{1}{r_1}\right),$$

and with R and S interchanged

$$\frac{S}{R} = P\left(\frac{1}{Q} + \frac{1}{r_2}\right).$$

Hence
$$\frac{R}{S} = \sqrt{\frac{1 + \frac{Q}{r_1}}{1 + \frac{Q}{r_2}}}.$$

See that the mercury contacts of the resistances are clean. Place the terminals of the two nearly equal coils, which are wound on the same bobbin, and those of the two coils to be compared in the mercury cups. Connect the galvanometer, and through a key the Leclanché cell, to the bridge, attending to rule (3), p. 278. Place a thermometer in the centre tube of each coil, and when the indications have become steady, read the temperatures. Obtain a balance and note the shunt resistance. Reverse the standard coils, balance again and note the reading. Read the thermometer again. Reverse the cell and balance. Read the temperatures.

Record as follows:

Date:

$Q = 1{\cdot}001$ ohms.

Wolff's standard 1 ohm (R) compared with Hartmann and Braun's standard 1 ohm (S).

Temperature coefficient of R said to be zero.

" " " $S = {\cdot}00020$.

S according to the certificate is correct at 18°·2.

Balance with R on left side of bridge ... $r_1 = 532{\cdot}2$.

" " S " " " " ... $r_2 = 2470{\cdot}6$.

$t = 15°{\cdot}15$.

Hence
$$\frac{R}{S} = 1{\cdot}00072.$$

Resistance of S at 15°·2 = 1 − ·00060 = ·99940 ohm,

∴ " of R at 15°·2 = 1·00012 ohms.

SECTION LIX

CHANGE OF RESISTANCE WITH TEMPERATURE

Apparatus required: *Two coils of wire in a tube which can be raised in temperature, Post Office bridge, voltaic cell, and mirror galvanometer.*

The electrical resistance of a wire of a pure metal increases rapidly with increase of temperature, while that of a wire of an alloy increases more slowly, and by a proper choice of the constituents may be made to remain nearly constant at ordinary temperatures.

To determine the change of resistance of a wire due to temperature, the wire may be wound round a sheet of mica and placed, along with a thermometer for indicating its temperature, in some insulating oil in a test tube surrounded by water which can be heated.

The apparatus supplied is constructed on this principle. It consists of a brass vessel, in which is placed a test tube containing two coils, one of copper and one of platinoid wire, immersed in petroleum and joined to the three screws on the wooden disc through which the tube passes in such a way that the middle screw is connected to one end of both coils, while the other ends of the coils are separately connected to the other two screws.

Fill the brass vessel full of tap water, and place the disc through which the test tube passes over the vessel, so that the tube is immersed in the water (Fig. 117). Connect the common terminal of the two coils to a Post Office resistance bridge in the usual way, and the other terminals to two of the screws of a three-way key, the third screw of which is connected to the other bridge terminal, and determine the resistance of each coil at the temperature of the bath.

Measure the resistance of the wires connecting the copper coil to the bridge, by clamping the two ends under one of the disc screws, and adjusting the bridge for a balance. Do the same for the connecting wires to the platinum coil, then screw up the wires as at the commencement, and repeat the determination of the resistance of each coil.

The galvanometer should be sufficiently sensitive to allow the thousandth part of an ohm to be estimated, and the temperature readings should be correct to ·05° C.

Raise the temperature of the water to about 20° C., and keep it at that temperature till the reading of the thermometer in the test tube is steady, then repeat the observations of the resistances of the two coils.

Take further observations at about 30°, 40°, 50°, 60°, then cool the water, and redetermine the resistances at the same temperatures (within about a degree) as previously.

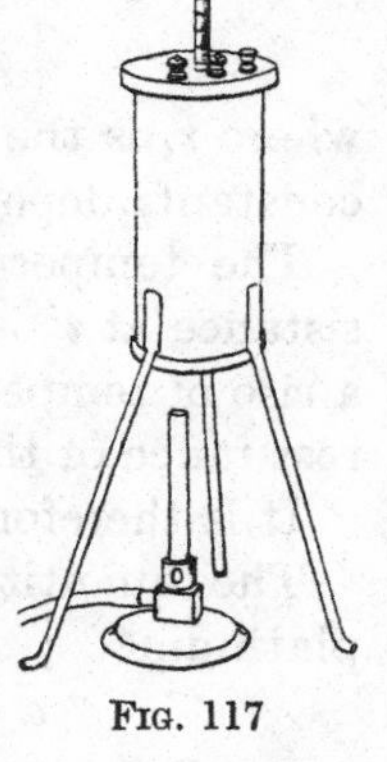

FIG. 117

Finally redetermine the resistance of the connecting wires.

Record for each coil as follows:

Date:

Resistance Box C. Coil No. 3.

Copper coil					
Temp. C.	Resistance ohms		Temp. C.	Resistance ohms	
	apparent	true		apparent	true
leads	·099		leads	·101	
13·62°	4·267	4·168	13·58°	4·263	4·162
20·15	4·373	4·274	20·05	4·359	4·258
29·80	4·518	4·418	29·90	4·534	4·434
39·15	4·652	4·552	39·35	4·684	4·584
49·50	4·827	4·727	49·60	4·843	4·743
59·90	5·015	4·915	59·70	4·981	4·881

Draw curves representing the results, taking temperatures as abscissae and resistances as ordinates.

It will be found that for the copper coil the curve is almost a straight line, which, when produced in the direction of temperatures below the freezing point, gives zero resistance approximately at the zero of absolute temperature.

Pure metals behave generally in the same way, but alloys do not.

The resistance of a pure metal may be represented very accurately by the equation

$$r_t = r_0 (1 + \alpha t + \beta t^2),$$

where r_t is the resistance at t° and r_0 at 0°, and α and β are constants depending on the nature and state of the metal.

The temperature coefficient of increase of electrical resistance at t° is the increase in resistance of a conductor for a rise of temperature from $(t - \frac{1}{2})^\circ$ to $(t + \frac{1}{2})^\circ$ divided by the resistance of the conductor at 0° C.

It is therefore in the above case $\alpha + 2\beta t$.

The quantity β is small compared to α: thus for pure platinum*

$$\alpha = + \cdot 00392, \quad \beta = - 0 \cdot 000{,}000585.$$

For many purposes it will be sufficient to neglect β, and to take α to be the mean temperature coefficient for the range of temperature considered. In that case, writing R_1 and R_2 for the resistances at the temperatures t_1 and t_2, we have

$$R_1 = r_0 (1 + \alpha t_1),$$
$$R_2 = r_0 (1 + \alpha t_2),$$

and by elimination of r_0

$$\alpha = \frac{R_2 - R_1}{R_1 t_2 - R_2 t_1} = \frac{\dfrac{1}{R_1} - \dfrac{1}{R_2}}{\dfrac{t_2}{R_2} - \dfrac{t_1}{R_1}}.$$

The first formula involves rather less calculation than the second, but the second allows a more symmetrical arrangement of the observations and calculations, and as tables of reciprocals should be available in every laboratory, the introduction of conductances $\frac{1}{R}$ in place of resistances involves little additional labour.

* Holborn, *Ann. der Phys.* VI, p. 251 (1901), Temperature on Nitrogen Scale.

Calculate the temperature coefficient of the electrical resistance of copper by taking the mean of each pair of nearly equal temperatures in the first and fourth columns of the above table and the mean of the resistances found for these temperatures. Arrange them in two sets and carry out the work as follows:

Date:

Copper. Coil 3.

Mean temp. Cent.	Mean resist. Ohms.	$\frac{1}{R}$	$\frac{t}{R}$	Mean temp. Cent.	Mean resist. Ohms.	$\frac{1}{R}$	$\frac{t}{R}$	Diff. of $\frac{1}{R}$	Diff. of $\frac{t}{R}$	Ratio
13·60°	4·165	·2401	3·265	39·25°	4·568	·2189	8·591	·0212	5·326	·00398
20·10	4·266	·2344	4·711	49·55	4·735	·2112	10·466	·0232	5·755	403
29·85	4·426	·2259	6·744	59·80	4·898	·2042	12·211	·0217	5·467	397
									a = mean =	·00399

Reduce the results for the platinoid coil in the same way and examine whether in its case the above simple relation between temperature and resistance holds.

Note on Platinum Thermometry. The measurement of change of resistance due to change of temperature has become of considerable importance, since it has been found that it may serve as a basis for the measurement of temperature. One of the advantages of this method is, that by the use of a metal with a high fusing point, *e.g.* platinum, it may be applied to temperatures which are so high that mercury or air thermometers cannot be used. A platinum thermometer must agree with the ordinary scale at the fundamental points 0° C. and 100° C., and this may be secured by defining the temperature as measured by the platinum thermometer (t_p) to be

$$\frac{t_p}{100} = \frac{r_t - r_0}{r_{100} - r_0},$$

where r_0, r_t, r_{100} are the resistances of a platinum wire at 0°, t° and 100° respectively. This definition makes the rise of temperature proportional to the increase in the resistance, and secures also that when r_t is equal to r_0 or r_{100} the temperature indicated shall be 0 or 100 respectively.

The difference between a platinum and a gas thermometer at any temperature can be calculated if the change of resistance of platinum as depending on the indications of the gas thermometer is known. Thus putting

$$r_t = r_0 (1 + \alpha t + \beta t^2),$$

where t is the temperature as measured on the gas thermometer, we find by substitution

$$t_p = \frac{\alpha t + \beta t^2}{\alpha + 100\beta},$$

and

$$t - t_p = \frac{10{,}000\beta}{\alpha + 100\beta} \cdot \frac{t}{100}\left(1 - \frac{t}{100}\right) = \delta . \frac{t}{100}\left(1 - \frac{t}{100}\right),$$

where $$\delta = \frac{10{,}000\beta}{\alpha + 100\beta},$$

an equation first used by Prof. Callendar. With the values of α and β given on p. 302, the difference of the two temperature scales becomes finally

$$t - t_p = 1 \cdot 51 \frac{t}{100}\left(\frac{t}{100} - 1\right).$$

The numerical coefficient has however to be redetermined for each sample of platinum, and appears to increase in value at very low temperatures.

SECTION LX

THE RESISTANCE OF ELECTROLYTES

Apparatus required: *Post Office resistance box, electrolyte tube, telephone, small induction coil, storage cell, and microscope.*

The resistance of an electrolyte cannot be measured by placing the vessel containing the liquid in the resistance bridge in the ordinary way, since the passage of the current produces polarisation at the electrodes, and therefore sets up an electromotive force, which has the same effect on the measuring instrument as a change of resistance. If however an alternating instead of a direct current is sent through the bridge, the polarisation due to the passage of the current in one direction is neutralised by the passage the next instant of an equal current in the opposite direction, and if the changes succeed each other with sufficient rapidity no appreciable effect on the measurement is produced. The galvanometer must however be replaced by an instrument capable of detecting alternating currents, as for example an electrodynamometer or a telephone. The latter is more generally used at present. The use of alternating currents necessitates care in avoiding appreciable self-induction and capacity in the resistances, since when they are present the resistance bridge does not measure "resistance" but "impedance," which depends on the self-induction, the capacity and the number of alternations per second as well as on the resistance. It is owing to the impedance depending on the rate of alternation that it is almost impossible to secure a perfect balance unless the alternating currents used follow the law of sines.

The concentration of a solution is expressed by the number of gram-equivalents of the dissolved salt present in 1 c.c. of the solution, and the dilution by the number of c.c. of solution which contain one gram-equivalent. By gram-equivalent is meant the molecular weight of the dissolved salt divided by its valency. It is thus equal to the molecular weight in the case of compounds like NaCl, HCl, $AgNO_3$, and to half the molecular weight for compounds like $ZnSO_4$, H_2SO_4, $BaCl_2$. A solution which contains one gram-equivalent of a salt per litre of solution, *i.e.* for which $\eta = \cdot 001$, is called a "normal solution." The "equivalent conductivity" K of an electrolyte is the conductivity divided by the number η of gram-equivalents dissolved in one cubic centimetre.

If it is required to measure the absolute value of the conductivity of an electrolyte, the electrolyte must be inclosed in a tube of known length (l) and cross section (a). If R is the resistance of such a tube Ra/l will be the resistivity, and l/aR the conductivity. If the tube is conical, a_1 and a_2 being the cross sections at the ends, it will have a resistance equal to that of a uniform tube having a cross section equal to $\sqrt{a_1 a_2}$. If each cross section is circular, its area is calculated in the usual way from the diameter, but if it is elliptical in shape, the two principal diameters d_1, d_2 must be measured, and the area is then equal to $\frac{1}{4}\pi d_1 d_2$. If d_3, d_4 represent the principal diameters at the other end, the cross section of the equivalent uniform tube will therefore be $\frac{1}{4}\pi\sqrt{d_1 d_2 d_3 d_4}$. In general the four diameters will be nearly equal, and if they do not differ by more than 2 or 3 per cent. it will be sufficient to substitute arithmetical for geometrical means, so that if $d = \frac{1}{4}(d_1 + d_2 + d_3 + d_4)$ the area to be used in the reduction of the experiment will be $\frac{1}{4}\pi d^2$.

Wash out the narrow glass tube and wider end tubes provided, and fill them with a solution of sodium chloride of four times normal strength, *i.e.* containing 4 gram-equivalents, *i.e.* 4 (23 + 35·5) = 234 grams, per litre of the solution, place the platinum electrodes in the end tubes and connect to the bridge as shewn in Fig. 118, taking out the 1000 ohm plugs from each arm of the box before connecting up. Place the bulb of a thermometer in one of the end tubes, taking care that it does not come between the platinum electrode and the end of the narrow tube. To the points C, D of the bridge to which the battery is usually connected, connect the terminals of the secondary of a small induction coil I, the primary of which is excited by a cell E of sufficient power to work the coil. Connect a telephone T to the terminals A, B.

Determine the resistance in the adjustable arm of the bridge to produce a minimum sound in the telephone and calculate S the resistance of the electrolyte.

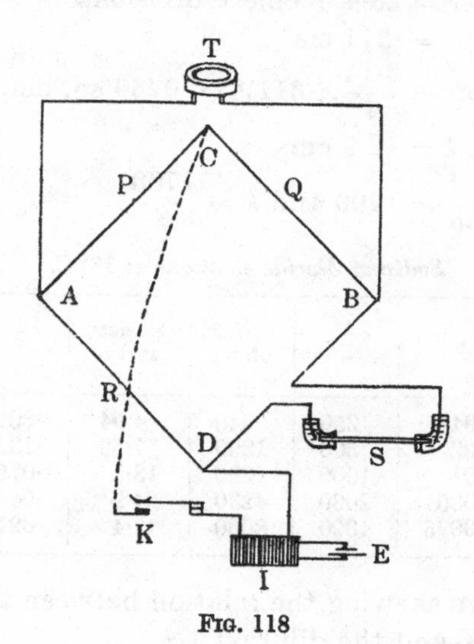

FIG. 118

Take about 50 c.c. of the solution and dilute to double the volume, *i.e.* make a solution of twice normal strength, and after washing out and drying the tube, fill it with the new solution and determine the resistance. Dilute down to normal, then to half, and quarter normal strength, determining the resistance and observing the temperature in each case.

Remove the end tubes of the electrolyte cell, and measure the internal diameter of each end of the centre tube under the microscope. If the tube is not quite circular measure the least and greatest diameters of each end. Measure the length l of the tube, and calculate the conductivity k of each solution.

Arrange your observations and results as follows:

Date:

Electrolyte tube A. Resistance box C. Microscope A.

75·4 eyepiece divisions = 1 cm. of stage scale,

∴ 1 „ division = ·0133 cm.

Diameters = 23·6, 23·2, 23·4, 23·4 eyepiece divisions.

Mean Diameter = 23·4 eyepiece divisions,

= ·311 cm.

$$\text{Area of section} = \frac{3 \cdot 14}{4} . (\cdot 311)^2 = \cdot 0759 \text{ sq. cm.}$$

Length of tube l = 12·9 cms.

$$\therefore \frac{l}{a} = 1700 \text{ and } k = \frac{1700}{S}.$$

Sodium chloride solutions at 18° C.

Strength	η	$\frac{1}{\eta}$	S ohms	Resistivity	k	Equivalent conductivity K
4 normal	·004	250	840	4·94	·202	50·5
2 „	·002	500	1290	7·59	·132	66
normal	·001	1000	2235	13·1	·076	76
½ normal	·0005	2000	4230	24·9	·040	80
¼ „	·00025	4000	8060	47·4	·021	84

Draw a curve shewing the relation between the equivalent conductivity K and the dilution $1/\eta$.

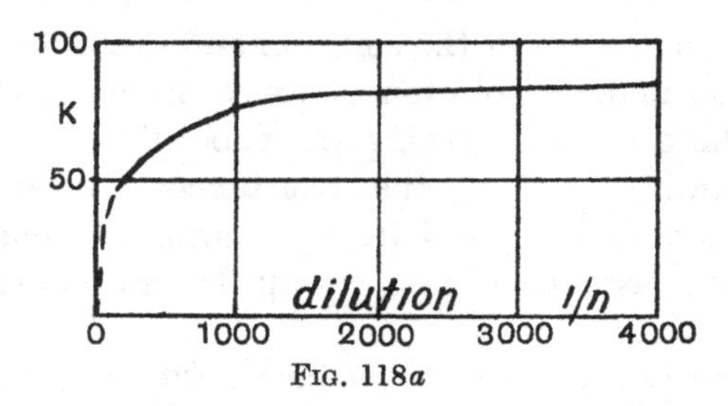

Fig. 118*a*

By placing the tube in a water bath, the influence of temperature on the conductivity of each electrolyte may be found, and expressed by a curve with temperatures as abscissae and conductivities as ordinates.

SECTION LXI

CONSTRUCTION OF A STANDARD CELL

Apparatus required: *H tube and stand, clean mercury, mercurous sulphate, pure cadmium, cadmium sulphate, reagents, paraffin and corks.*

In one form of the Weston standard cell the active materials are enclosed in a H-shaped tube (Fig. 119), through the lowest

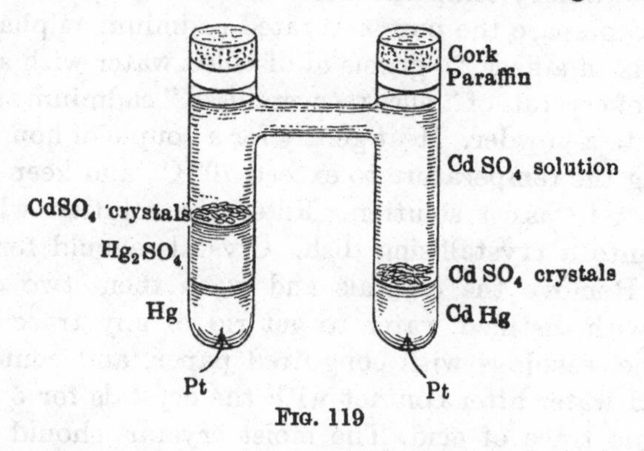

Fig. 119

points of which the platinum wires forming the terminals of the cell pass.

The wire on the left (Fig. 119) ends within the tube in a small quantity of pure mercury, and that on the right in cadmium amalgam. The mercury is covered with a paste of mercurous sulphate, and this again with a few crystals of cadmium sulphate. The amalgam is also covered with crystals, and the rest of the tube filled with a saturated solution of pure cadmium sulphate. Both tubes are closed by cemented corks.

In constructing a cell the following instructions, taken from the memorandum of the Board of Trade, should be carefully attended to.

1. To secure purity of the mercury it should be first shaken up in a bottle with dilute nitric acid (1 : 6), washed and then distilled *in vacuo*.

2. To prepare the cadmium amalgam add 2 grams of "commercially pure" cadmium to 18 grams of the mercury, and heat to 100° C. in an evaporating dish on a water bath. If the surface of the cadmium is clean it quickly becomes amalgamated, and the cadmium slowly dissolves in the mercury, which should be occasionally stirred. If the amalgamation does not occur immediately, the cadmium should be removed, treated with dilute sulphuric acid, washed, dried and replaced. The amalgam should be liquid at 60° C. and solid at ordinary temperatures.

3. To prepare the pure saturated cadmium sulphate solution, mix in a flask 20 grams of distilled water with an equal weight of crystals of "pure recrystallised" cadmium sulphate, ground to a powder. Heat gently for a couple of hours, never allowing the temperature to exceed 70° C., and keep shaking the flask to assist solution. Filter the solution while still warm into a crystallising dish. Crystals should form as it cools. Remove the crystals and wash them two or three times with distilled water to get rid of any trace of acid. Test the washings with congo red paper, and continue till distilled water after contact with the crystals for 5 minutes shews no trace of acid. The moist crystals should then be transferred to a stock bottle as acid free $CdSO_4\ 8/3\ H_2O$.

At temperatures above 70° C. the salt may crystallise out in another form; to avoid this 70° C. should be the utmost limit of temperature. At this temperature water dissolves about ·8 times its weight of the crystals. If any crystals remain undissolved they are removed by the filtration. To prepare the saturated solution of cadmium sulphate for the cell take about 10 grams of the acid free crystals in an equal weight of distilled water. Heat to 45° C. in a flask and shake well. About ·8 of the salt should enter into solution.

4. Take about 10 grams of mercurous sulphate Hg_2SO_4 purchased as pure, and wash it thoroughly with dilute sulphuric acid (1 : 6), by agitation in a bottle with a few drops

of mercury; drain off the acid and wash with distilled water several times till the washing water no longer shews signs of acidity when tested with congo red paper. After the last washing drain off as much of the water as possible. Mix the washed mercurous sulphate with a few drops of the acid free cadmium sulphate solution, and about half its weight of pure mercury, adding sufficient crystals of cadmium sulphate from the stock bottle to ensure saturation. Rub these well up together in a mortar to form a paste of the consistency of cream. This ensures the formation of a saturated solution of cadmium and mercurous sulphates in water.

The above treatment of the mercurous sulphate has for its object the removal of any mercuric sulphate, which may be present as an impurity. Mercuric sulphate decomposes in the presence of water into an acid and a basic sulphate. The latter is a yellow substance—turpeth mineral—practically insoluble in water; its presence, at any rate in moderate quantities, has no evil effect. If, however, it is formed, the acid sulphate is also formed. This is soluble in water and the acid produced affects the electromotive force of the cell. The object of the washings is to dissolve and remove the acid sulphate, and for this purpose the washings described will in nearly all cases suffice. If, however, a great deal of turpeth mineral is formed, it shews that there is a great deal of the acid sulphate present, and it will then be wiser to obtain a fresh sample of mercurous sulphate rather than to try by repeated washings to get rid of all the acid. The free mercury helps in the process of decomposing the acid salt, forming mercurous sulphate and sulphuric acid, which will be washed away.

The materials having been prepared, pure mercury and cadmium amalgam are respectively poured into the two vertical parts of the H tube till the platinum wires are covered. The amalgam, which is solid at ordinary temperatures, should be heated till it is liquid, and the limb of the H tube intended to contain it heated to about the same temperature. It should then be poured into the H tube down a hot glass tube of outside diameter less than the inside diameter of the H tube, to prevent its soiling the sides of the H tube. The mercurous sulphate paste should then be forced down a glass tube on to the mercury, and the tube withdrawn, care being taken not to soil the H tube. A few crystals of cadmium sulphate should then be placed on the surfaces of the paste and amalgam, and the rest of the tube up to about 1·5 cms. from the top filled with the concentrated cadmium sulphate solution.

A small quantity of clean paraffin wax should then be melted, the tube tilted a little to the left and the paraffin poured gently on to the surface of the solution in the left-hand tube, till a layer about half a centimetre thick is formed. The tube should then be tilted to the right and the right-hand tube filled in the same way. This process secures a layer of air between the liquid in each tube and the paraffin, which will admit of increase of volume of the liquid owing to rise of temperature without the tube being broken. On the top of the layers of paraffin, corks about ·5 cm. thick should be placed, and the tops of the tubes then sealed with marine glue.

A label bearing the date and the name of the maker should be attached to the stand on which the tube is supported.

The cell should stand a few days and then be compared with a standard cell by the method described in the next exercise, and the result recorded on the label.

Instead of the two limbs of the H tube being sealed with cement they may be drawn out in the blowpipe flame and hermetically sealed.

If the cell is to be portable the two limbs may be slightly constricted about 2 cms. above their lowest points, and the quantities of amalgam of mercury and mercurous sulphate so adjusted that the tops of the layers of crystals of cadmium sulphate are in the constrictions and form plugs keeping the materials below them in place.

The Electromotive Force of a Weston cell constructed in this way has at $t°$ C. the value $E_t = 1{\cdot}0183 - {\cdot}00004\,(t° - 20°)$ volts very nearly.

SECTION LXII

ELECTROMOTIVE FORCES

Apparatus required: *Two similar resistance boxes, high resistance, mirror galvanometer, Daniell, Leclanché, and Standard cells, and connecting wires.*

When the electromotive forces of different cells are to be compared together it is necessary to carry out the comparison under standard conditions, and the condition usually adopted is that the cell under test shall be given no current, or only an extremely small one, at the time of the test. The electromotive force E_1 of the cell must therefore be balanced, the balance being indicated by a galvanometer in series with the cell remaining undeflected. The balancing electromotive force is best provided by the difference of potential between two points of a resistance through which a current is passing, and wires from the cell under test are brought through a galvanometer to these two points. If R_1 is the resistance between the points of contact, C the current through that resistance, and E_1 the difference of potential between the points, we have

$$E_1 = CR_1.$$

If a second cell of electromotive force E_2 is substituted, and balance exists when the resistance between the points of contact is R_2, we have, if the current is the same,

$$E_2 = CR_2.$$

Hence

$$\frac{E_2}{E_1} = \frac{R_2}{R_1},$$

which gives the ratio of the electromotive forces.

Connect two similar resistance boxes A and B (Fig. 120), each of about 10,000 ohms, in series with a plug-key K and two Leclanché cells, taking out the plug from the key before making connections. Set up a Daniell cell, with a clean zinc plate in zinc sulphate, and a clean copper plate in copper sulphate. To the terminals of one of the boxes, say A, connect, in series with each other, the Daniell cell, a galvanometer G, a resistance R of about 100,000 ohms, and a spring key K_2, arranging that the cell under test, if alone, would send a current through the box A in the same direction as that sent by the Leclanchés.

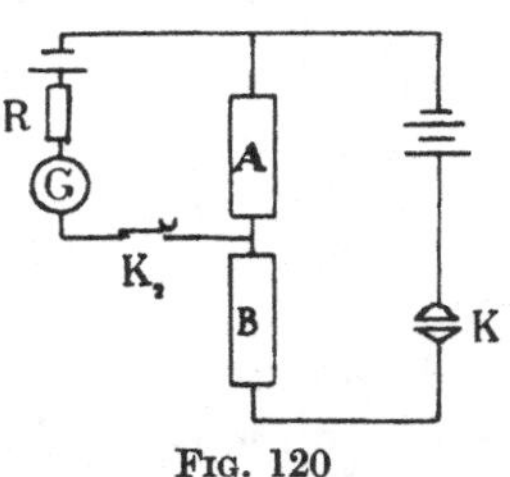

FIG. 120

Take out plugs for 10,000 ohms from the box A. Insert the plug in the key K, make connection for an instant at the spring key K_2 in the test cell circuit, and observe the direction in which the spot of light moves. Take out plugs for 5000 ohms from the box B, plugging 5000 ohms in A so that the sum of the resistances of A and B remains the same. Make contact at the key K_2, and notice the direction of motion of the spot. If it is the same as previously, take out more plugs from B and insert the same number in A. Continue adjusting the resistance of the two boxes, keeping the sum constant, till on closing the galvanometer circuit there is no deflection. If the arrangement is not sufficiently sensitive to enable the correct resistance to be found to within 1 ohm, determine it as nearly as possible, then short circuit the 100,000 ohms in the galvanometer circuit, and determine it more accurately. Make a note of the resistance in each box.

Insert again the 100,000 ohms in the galvanometer circuit, then replace the Daniell by a Leclanché cell and balance as before.

Then substitute the Standard cell, reading the temperature of the air in its neighbourhood, or better still, placing the cell in a water or oil bath, the temperature of which is measured. Substitute again the Leclanché and then the Daniell cell. Since a Standard cell made according to the instructions contained in Section LXI has a difference of potential at its electrodes of $1{\cdot}0183 - {\cdot}00004\,(t - 20°)$ volts, where t is the temperature Centigrade, the electromotive force of each of the other cells can be found in volts.

Calculate the electromotive force of the Daniell and of the Leclanché cells in volts.

Arrange observations and results as follows:

Date:

Resistance Boxes A and B.

Cell	Resistance A	Resistance B	Sum
Daniell, No. 4	3,248	6,752	10,000
Leclanché, No. 13	4,124	5,876	,,
Weston, No. 3, 18° C.	2,952	7,048	,,
Leclanché, No. 13	4,122	5,878	,,
Daniell, No. 4	3,246	6,754	,,

electromotive force of Weston at 18° = 1·018 volts.

$$\therefore \text{ electromotive force of Daniell} = 1{\cdot}02 \times \frac{3247}{2952} = 1{\cdot}10 \quad ,,$$

$$,, \qquad ,, \quad \text{Leclanché} = 1{\cdot}02 \times \frac{4123}{2952} = 1{\cdot}40 \quad ,,$$

The effect of temperature on the Daniell and Leclanché is masked by other irregularities, hence their temperatures need not be noted, and an accuracy of one per cent. in the result is sufficient, although the method will evidently give results correct to 4 figures.

A compact form of the apparatus used in this exercise is known as a "Potentiometer." In it the resistance A consists of 15 equal resistances in series, the one at the B end being a wire along which the contact to the galvanometer slides. The contact from the cell to be tested is made by a switch moving over the 15 contact pegs at the ends of the 14 equal resistances. The arrangement is shewn in a diagrammatic form in Fig. 121. The resistance $A + B$ may be too small compared to that of a Leclanché cell to admit of the use of such cells to provide the current, and a small storage cell is generally used. The rough adjustment for a balance is made by means of the switch and the fine adjustment by means of the slider on the wire.

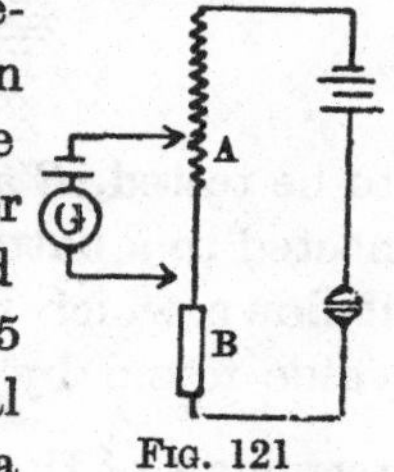

Fig. 121

If the main current is supplied by two Leclanchés, the original method may be used to measure any electromotive force less than about three volts, *e.g.*, that at the terminals of a voltmeter intended for testing storage cells.

When however the electromotive force to be measured exceeds a few volts, as, *e.g.*, when a voltmeter reading to 100 volts is to be standardised, the arrangement requires modifying slightly because the electromotive force of the battery supplying the current cannot be raised above about three volts for fear of overheating the resistances A and B.

The fall of potential between the terminals of the voltmeter is subdivided by connecting to the terminals a high resistance divided into a number of parts the resistances of which bear known ratios to that of the whole. The arrangement is known as a Volt Ratio Resistance Box. The difference of potential at the ends of one of these parts is then compared with that of the Weston cell by the above method. The circuit is arranged as shewn in Fig. 122, where V is the voltmeter

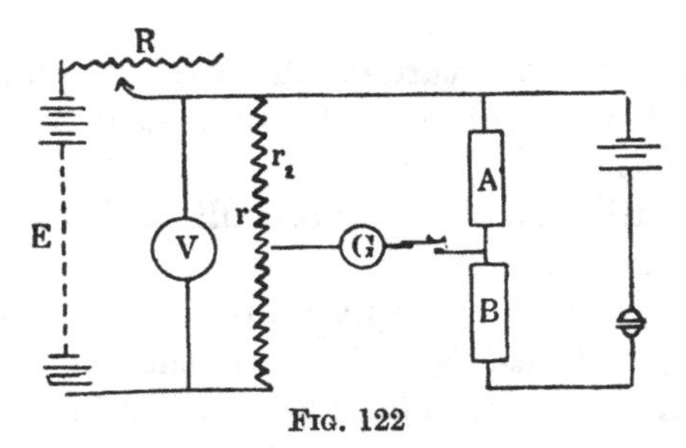

FIG. 122

to be tested, R a variable resistance through which it is connected to a battery E, r the high resistance, r_2 the portion of it down which the fall of potential is measured. If e is the value found by experiment, the electromotive force at the terminals of the voltmeter $= \dfrac{r}{r_2} e$.

Determine the electromotive force of the electric light supply by this method. Record as follows:

Date:

Volt Ratio Box Tinsley 4258. Ratio 1/200.

Rayleigh Boxes A and B = 11,111 ohms.

Balance on Ratio Box A = 6,050

" " Weston cell 21 = 5,025

$$\therefore \text{E.M.F. of mains} = \frac{6050}{5025} \times 1{\cdot}018 \times 200 = 240 \text{ volts.}$$

SECTION LXIII

THE POTENTIOMETER METHOD OF MEASURING CURRENTS

Apparatus required: ***Standard low resistance, adjustable resistance, storage cells, current measuring instrument to be standardised, Weston cell and mirror galvanometer.***

When an electric current of A amperes is sent through a resistance of R ohms, it creates a difference of potential of V volts between the ends of the resistance, where $V = AR$, and sometimes it is more convenient to measure the current by means of the difference of potential it produces at the ends of a known resistance, than to measure it directly. This method is known as the "potentiometer method." In order that the method may give accurate results, the resistance R should be made of a material having a small temperature coefficient, and should have a sufficiently large surface to prevent the temperature rising more than a few degrees.

We shall shew how the method is used to standardise a current meter.

Connect the current measuring instrument A, to be tested,

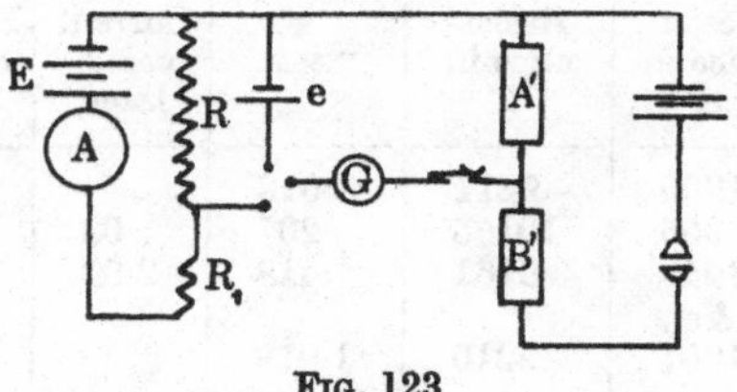

FIG. 123

to a standard low resistance R, capable of carrying the current which it is proposed to use without undue heating, and to an adjustable resistance R_1, using a number of storage cells E to supply the current required (Fig. 123).

Connect the ends of the resistance through a mirror galvanometer and tapping key to the ends of one of two similar resistance boxes A', B' arranged in series, and forming a circuit with two Leclanché cells. Find, as described on p. 314, the plugs which must be taken out of the two boxes to enable contact at the key to be made without a deflection of the galvanometer resulting. Now substitute for the low resistance R in the galvanometer circuit a Standard cell e, and again balance by adjusting the resistances in the two boxes, keeping their sum constant. If a reliable "Potentiometer" is available, it may be used instead of the two resistance boxes to compare the electromotive forces as explained on p. 315.

Observe the temperature of the Standard cell and calculate its electromotive force at that temperature. From the two values of the resistances in the boxes A and B calculate the electromotive force E at the ends of the standard resistance. If the resistance of the standard is R ohms, the current through the resistance and current measuring instrument is E/R amperes, and this should be compared with the current as registered by the instrument.

Record as follows:

Date:

Standardisation of Ammeter No. 4.

Standard low resistance C = ·20 ohm.

Weston cell No. 4 at 19° = 1·018 volts.

Connection to	Resistance in A'	Resistance in B'	E.M.F.	Current calculated	Reading on ammeter	Correction
Weston cell	1900	9211	1·018			
Ammeter	956	10155	·207	1·03	1·01	+ ·02
"	1940	9161	·418	2·09	2·05	+ ·04
	&c.					
Weston cell	1901	9210	1·018			

SECTION LXIV

THERMO-ELECTRIC CIRCUITS

Apparatus required: *Water baths, thermo-circuits, four-way key and mirror galvanometer.*

If a circuit consists of wires of different materials, and if one of the junctions of two dissimilar wires is heated, an electric current flows through the circuit, and continues to flow so long as the difference of temperature between the heated junction and the rest of the circuit is maintained.

This electric current is due to an electromotive force in the circuit produced by the inequality of temperature of the two junctions, and it is found, for small differences of temperature between the two junctions, to be nearly proportional to the difference. For greater differences, if t_1 is the temperature of the hot junction, t_0 that of the rest of the circuit, the electromotive force e in a circuit, of two metals is given by the equation

$$e = A\,(t_1 - t_0)\left(T - \frac{t_1 + t_0}{2}\right),$$

where A and T are constants depending on the two materials of the circuit, T being a temperature known as their "neutral temperature."

To verify the above statements the apparatus shewn in Fig. 124 is provided.

It consists of two vessels containing water, in which are placed two test tubes containing the junction of the wires to be experimented on, and thermometers for indicating their temperatures. The rest of each tube is filled with clean sand or with petroleum, to improve the thermal connection of the junctions and thermometers with the water.

The circuits to be tested consist of lengths of No. 25 iron, nickel, and lead wire, to one end of each of which a length of copper wire is soldered, and brought to a binding screw placed on the board through which the test tubes pass. The other ends are soldered to copper wire brought to a fourth screw on the board.

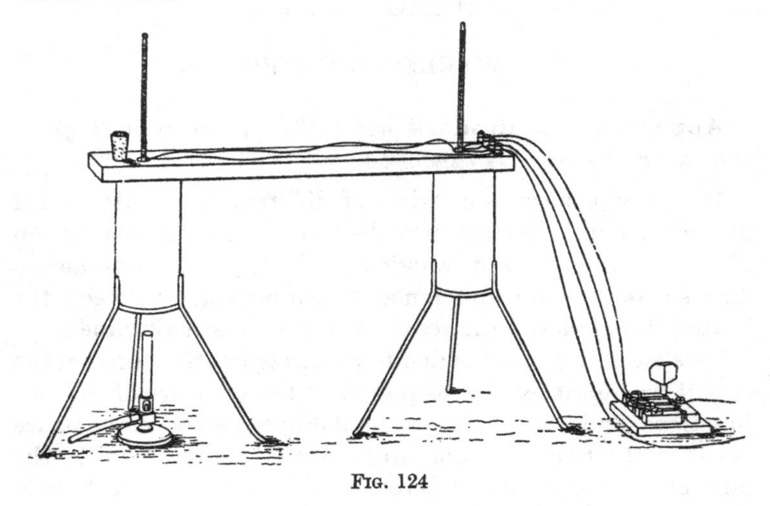

Fig. 124

The binding screws except that connected with the lead wire should be joined by copper connecting wires to a four-way key, so that each wire may in turn be connected together with the lead wire to a galvanometer of about 50 ohms resistance. With a galvanometer of this resistance the effect of the different resistances of the circuits may be neglected, and the deflections taken as proportional to the electromotive forces acting in the various circuits.

Fill the two vessels with water at the temperature of the room, and connect the thermo-circuits in turn through the four-way key to the galvanometer. Verify that there is no current in any of the circuits.

Now raise the temperature of the vessel, which has no binding screws over it, to about 70° C. and keep it constant for 10 minutes. Then connect the galvanometer to each circuit in turn, and determine the deflections, noting the temperature before and after each observation. If the sensitiveness of the galvanometer can be altered adjust it so that the greatest deflection observed is nearly to the end of the scale. Decrease the temperature to about 60° C. by adding cold water and repeat.

Continue till the hotter vessel reaches about 20° C., then raise its temperature to about 70° C. in steps of about 10° C., taking observations in the same way during the process.

Note the direction in which the current flows through the hot junction in each case, and enter the deflection as positive when the current flows from the lead to the other metal through the hot junction.

Record as follows:

Date:
Apparatus *C*.

t_0	t_1	Deflections		
		Lead iron	Lead copper	Lead nickel
17·1	70·0	+ 7·2	+ 1·85	− 18·0
17·1	59·0	5·5	1·30	− 13·3
17·1	49·0	4·5	1·00	− 9·7
	&c.	&c.	&c.	&c.
17·0	28·5	1·5	·35	− 3·0
17·0	39·5	3·0	·65	− 6·3
17·1	52·0	4·7	1·20	− 10·7
	&c.	&c.	&c.	&c.

Represent the observations for each circuit by a curve taking $t_1 - t_0$ as abscissae and the deflections as ordinates. Indicate the observations taken while the temperature was decreasing by a cross, those while it was increasing by a circle. The curves will be found to be almost straight lines, and from this it is evident that if we put the equation (p. 319) into the form

$$e = E_0 (t_1 - t_0)(1 - b\overline{t_1 + t_0}), \text{ where } E_0 = AT \text{ and } b = \frac{A}{2E_0},$$

b must be small.

The "thermo-electric power or height" of a metal at a given temperature with respect to lead, which is taken as the standard, is defined as the ratio of the small increase of E.M.F. produced when a junction of lead and the metal is heated, to the small increase of temperature of the junction, and is counted positive if the E.M.F. tends to produce a current from the lead to the metal through the heated junction.

In the above case, if the deflection obtained say for lead-nickel at 59° is subtracted from that obtained at 70° and the difference divided by 70° − 59°, the quotient is proportional to the thermo-electric power of nickel at $(70 + 59)/2 = 64{\cdot}5°$ C. on the scale used. To reduce the results to an absolute scale, in micro-volts per degree, make use of the fact that the thermo-electric power of iron with respect to lead at about 100° C. is 1 micro-volt per degree.

Determine from the observations the thermo-electric powers of the metals used, and draw a "thermo-electric diagram," taking temperatures as abscissae and thermo-electric powers as ordinates.

Arrange as follows:

Lead Nickel.

t_1	Difference	Mean	Deflection cms.	Difference cms.	Ratio of differences	Thermo-electric power
17°·1			0			
28 ·5	11°·4	22°·8	− 3·0	− 3·0	− ·265	− 1·56
39 ·5	11 ·0	34 ·0	− 6·3	− 3·3	− ·300	− 1·75
&c.			&c.			&c.
59 ·0			− 13·3			
70 ·3	11 ·0	64 ·5	− 18·0	− 4·7	− ·423	− 2·50

Similarly for the other metals.

If more accurate results are required, a low resistance galvanometer must be used, and the resistance of each circuit be made equal, or the total resistance of the galvanometer and each circuit be taken into account in comparing the electromotive forces of the circuits.

If the electromotive forces of the junctions are required in absolute measure balance them against the fall of potential down a small resistance of say ·1 ohm in series with a large resistance of the order 10,000 ohms, and a Daniell cell of E.M.F. 1·07 volts.

SECTION LXV

THE MECHANICAL EQUIVALENT OF HEAT BY THE ELECTRICAL METHOD

Apparatus required: *Covered calorimeter, thermometers, heating coil, standardised ammeter and voltmeter, storage cells, watch.*

When the whole of the work done on a body is converted into heat, the amount of work done bears a fixed ratio to the amount of heat produced, in whatever way the work is performed, and the work which has to be done to generate one gram-degree of heat is, we have seen (p. 139), known as the "mechanical equivalent" of heat. To determine this quantity, any convenient method of generating heat by performing work on the body may be adopted, and it is proposed in this section to do work by sending a current of electricity through an insulated wire immersed in water.

If A is the current passing through the wire, and E the electromotive force at the ends of the wire, the rate at which work is done on the wire per second is EA watts, and if the current flows for t seconds, the total work done $= EAt$ joules $= EAt \cdot 10^7$ ergs. If the water rises in temperature θ degrees, and the water equivalent of the calorimeter thermometer and coil $= w$, the heat generated, supposing no heat is lost by radiation &c. $= w\theta$. If J joules are necessary to generate one gram-degree of heat, we have

$$EAt = w\theta J,$$

from which J can be found.

Weigh the calorimeter and stirrer provided. Nearly fill the calorimeter with water and weigh again. Weigh also the platinoid resistance coil, and support it from the wooden lid of the calorimeter, taking care that it does not touch the sides. Place a thermometer graduated in tenths of degrees in the water. Connect the coil through a standardised ammeter and an adjustable resistance to sufficient storage cells to furnish the current required. Connect a standardised voltmeter of known resistance to the ends of the heating coil. The resistance of the voltmeter is required in order that the current through it may be calculated and subtracted from that indicated by the ammeter. Make circuit and see that the instruments give proper indications and that the thermometer shews a gradual rise of temperature. Break the circuit, stir the water well, and after a few minutes take observations of temperature every half minute as described on p. 125. At the end of the first period put on the current, read the thermometer every half minute, observing the voltmeter 15 seconds before each minute and the ammeter 15 seconds after each minute. This second period should continue till the temperature of the calorimeter has risen about three degrees, then at the end of one of the intervals the current should be switched off and observation of temperature continued till the rate of change is steady. From the first and third periods the cooling corrections during the second and third periods should be calculated as in pp. 125–126, and from the initial and final corrected temperatures the rise of temperature determined.

J is calculated by substituting in the above equation for E the mean electromotive force, and for A the mean current, if both quantities shew only small variations during the experiment. If they are not sufficiently constant, the product of EA must be calculated for each interval of time and the mean product substituted in the equation.

Record as follows:

Date:

Weight of calorimeter	= 55·0	grams
,, ,, and water ...	= 245·2	,,
,, water	= 190·2	,,
,, platinoid coil	= 28	,,
Water equivalent of calorimeter and coil	= 7·8	,,

Total water equivalent	... ...	= 198 grams
Rise of temperature ...	... ...	= 3°·11 C.
∴ Heat generated ...	... ...	= 616 gram-degrees
Mean electromotive force*	... ...	= 1·45 volts
Resistance of voltmeter	... ...	= 14 ohms
Mean current through ammeter	...	= 11·10 amperes
" " "	voltmeter = $\frac{1 \cdot 45}{14}$	= ·10 "
" " "	coil	= 11·00 "
Time	...	= 161 seconds
Work done = 1·45 × 11 × 161	...	= 2575 joules

$$\therefore \text{Equivalent} = \tfrac{2575}{616} = 4 \cdot 18 \text{ joules per gram-degree}$$

$$= 41 \cdot 8 \times 10^6 \text{ ergs per gram-degree}$$

This value happens to be almost exactly right, but errors of one per cent. are likely to occur unless the voltmeter and ammeter have been carefully standardised.

* The resistance of the heating coil is made small so that the difference of potential between its ends may be small enough to prevent electrolysis of the water. The electrolysis difficulty may be obviated by using a non-conducting oil whose specific heat is known in place of water, or by winding the heating coil on the outside of but insulated from the calorimeter.

SECTION LXVI

INDUCTION OF ELECTRIC CURRENTS

Apparatus required: *Two solenoids of known resistance, one sliding within the other, tangent galvanometer or ammeter reading* 1 *ampere, reversing switch, mirror galvanometer of low known resistance, resistance coils.*

When a current is made, broken, or altered in strength in any circuit, induced currents are produced in neighbouring circuits, and it is the object of this exercise to find on what conditions the magnitudes of these induced currents depend.

The induced currents will last only a very short time, and a galvanometer in one of these neighbouring circuits will not measure the strength of the current, but the total quantity of electricity which has passed through it. If α is the angle of the first swing of the galvanometer needle produced, it is shewn (p. 331) that the quantity of electricity which has passed through the coil of the galvanometer is proportional to $\sin\frac{\alpha}{2}$. If the galvanometer needle hangs in its proper position when no current passes through the instrument, and the scale is properly adjusted,

$$\sin\frac{\alpha}{2} = \frac{x}{4d}\left\{1 - \frac{11}{32}\left(\frac{x}{d}\right)^2\right\},$$

where x is the observed deflection and d the distance of the scale from the mirror (p. 156). If an error not exceeding half per cent. is allowable, and the deflections never exceed 25 cms. on a scale placed a metre from the mirror, the second term on the right-hand side is negligible; and we may therefore take the observed reading x to be proportional to the quantity of electricity which has passed through the galvanometer. The experiments of the present exercise are supposed to be made under these conditions.

Two solenoids, *P*, *S* (Fig. 125), mounted on blocks of wood so that their axes are coincident, are provided. The inner coil *P* can be moved in a direction parallel to the axis, and placed with its centre at any convenient distance from the centre of the outer fixed solenoid *S*. Each solenoid is divided into three parts, and the number of turns to each part should be counted and recorded.

Fig. 125

Arrange the turns of the inner coil in series with each other and with a cell, a reversing switch, an adjustable resistance, and a tangent galvanometer or ammeter. Place it within the outer coil so that the centres of the coils coincide, the mark on its base will then read 0 on the scale of the outer coil. Connect, by means of copper wire (about no. 18), the end terminals of the outer coil through a resistance box to a low resistance mirror galvanometer. At first cut the mirror galvanometer out of circuit by connecting the two wires leading to it to the same terminal, and observe whether making or breaking the battery circuit has any effect on the galvanometer*. If so remove the coils further away and place their axis in such a direction that this is no longer the case, wherever the inner coil is placed on its slide.

Adjust the resistance in series with the inner coil till the current flowing through it is one ampere, and take readings occasionally to see that it remains steady. Notice that making or breaking the battery circuit produces a deflection of the mirror galvanometer needle. Observe the extent of the first swing and verify that the swing on making is equal and

* This is generally unnecessary if the galvanometer is a moving coil instrument.

opposite to that on breaking the primary, *i.e.* the battery circuit, and that the swing on reversing the primary current is double the previous swings. If this is found not to be the case, cut the galvanometer out of circuit as before and make sure that there is no direct action from the primary coil.

Effect of the relative positions of the two coils. Arrange the resistances so that on making or breaking one ampere in the primary, the deflection is about half the greatest observable deflection. Determine its amount, then slide the inner coil through 1 cm. in a direction parallel to the axis of the coils, and again determine the deflection. Repeat for 2, 5, 10, 15 and 20 cms., thus gradually sliding the inner coil out of the outer one.

Place the inner coil on the table, move the outer coil to a distance of about 50 cms., and place it with its centre on the axis of the inner coil produced. Find the direction in which its axis must point in order that there is no deflection on making or breaking the primary circuit.

Move the inner coil towards the outer, keeping it parallel to itself, and verify that it may be adjusted so that there is no deflection even when the coils are near together.

Verify this for positions in which the centre of the outer coil is not on the axis of the inner coil.

Effect of the resistance of the secondary circuit. Replace both coils with the inner coil in the position in which the induced current was found to be a maximum, and determine the swing on making or breaking the primary circuit.

Double the total resistance in the mirror galvanometer (or secondary) circuit by taking out plugs from the box equal to the sum of the resistances of the galvanometer, coil and connecting wires, and verify that the swing is now half what it was before.

Effect of the magnitude of the current in the primary. Increase the resistance in the battery circuit till the current has half its original value, and verify that the swing on making or breaking is half what it previously was.

Effect of number of turns in the primary. Adjust the resistance till the primary current is again one ampere. Change the connections of the primary so that the current passes through two coils only. The connections are arranged so that the resistance in circuit, and therefore the current, remains the same; but if any variation is observed, adjust the resistance in series with the storage cell till the current is again one ampere, and observe the swing on making or breaking. Notice that it is decreased in the ratio 3 : 2. Now connect the cell to the two centre screws so as to send the current through the central coil only. Notice whether the current remains the same; if so the deflections will be found to be again decreased in the ratio 2 : 1.

Effect of number of turns in the secondary. Returning to the whole of the turns on the primary, change the connection of the galvanometer from 3 to 2 coils of the secondary. The terminals of the coils are so arranged that the resistance in circuit remains the same. Verify that on now making or breaking the primary circuit, the swing is decreased in the ratio 3 : 2. Make connections to the centre coil only, the arrangement again secures that the resistance remains constant. The deflection will be found to be decreased in the ratio 2 : 1.

It has therefore been shewn that for a given position of the coils with respect to each other, the induced current in the secondary is proportional to the current in the primary, to the number of turns of the primary, to the number of turns of the secondary, and inversely proportional to the resistance of the secondary circuit.

Effect of moving magnets. Instructive experiments may be made, when two observers are available, by moving magnets in the neighbourhood of the secondary coil. For this purpose the outer coil should be removed to a distance such that the bar magnet to be used in the experiments has no direct effect on the galvanometer.

Place the magnet in the centre of the coil, let one observer remove it quickly to a distance from the coil, and let the other observer note the deflection produced.

Replace the magnet in its original position and withdraw it again, but pulling it out towards the other side. This should produce an equal deflection *and in the same direction.*

Place the magnet outside the coil and in such a position that its axis coincides with that of the coil; note the deflection when it is suddenly moved to a distance, and shew that it is the same as if the magnet were turned through a right angle about an axis through its centre perpendicular to its own axis.

Verify that when a magnet is moved quickly from a position A to a position B the effect on the galvanometer is the same through whatever path the magnet is moved. Also try to shew that when the magnet can be moved from a position A along any path back to the same position A in a time which is short compared to the time of oscillation of the galvanometer needle, the total effect on the latter is nil.

SECTION LXVII

STANDARDISATION OF A BALLISTIC GALVANOMETER

Apparatus required: *Ballistic galvanometer, megohm, Daniell cell, solenoid and small coil, storage cells, reversing switch and ammeter.*

(a) **Steady Deflection Method.**

If the galvanometer is of the suspended needle type with the steady external field H and the field produced at the centre of the coil when unit current flows round it G, and the quantity Q discharged through it gives an angular throw α not exceeding 8°, then

$$Q = \frac{H + \dfrac{N}{M}}{G} \cdot \frac{T}{\pi} \sin \frac{\alpha}{2},$$

where T is the time of swing and M the magnetic moment of the needle, and N the torsional constant of the suspension (the couple to deflect it a radian). N/M is generally small and usually neglected.

The value of H/G can be determined experimentally by observing the permanent deflection θ of the galvanometer, produced by a known current C, for in that case if θ does not exceed 8°

$$GC = \left(H + \frac{N}{M}\right) \tan \theta$$

provided that the galvanometer needle moves in a field which is sufficiently uniform for the tangent law to be true. Otherwise $\tan \theta$ must be replaced by some function of θ, obtained by calibrating the instrument.

If the simpler supposition is sufficiently accurate,

$$Q = \frac{T}{\pi} \frac{C \sin \dfrac{\alpha}{2}}{\tan \theta}.$$

If the galvanometer be of the suspended coil type, A the equivalent area of the coil, H' the mean field assumed radial, in which the coil rotates, and the quantity Q discharged through it gives an angular throw α,

$$Q = \frac{N}{AH'} \cdot \frac{T}{2\pi} \cdot \alpha.$$

If a steady current C passing through the instrument causes a deflection θ we have

$$AH'C = N\theta,$$

and
$$Q = \frac{T}{2\pi} \cdot \frac{C\alpha}{\theta} * .$$

Exercise I. Standardise a galvanometer by this method, using as steady current that given by a Daniell cell in series with a megohm and the instrument. It will be sufficient to assume that the electromotive force of the Daniell is 1·07 volts, but for more accurate work this electromotive force should be determined in terms of a standard Weston cell. The resistance G of the galvanometer is assumed known. The current C through the instrument is $1{\cdot}07/(10^6 + G)$ amperes. Observe the deflection D due to this current.

Then if d_1 is the extent of the first throw when a quantity Q is discharged through the galvanometer,

$$\frac{\sin \frac{\alpha}{2}}{\tan \theta} = \frac{d_1}{2D} \text{ or } \frac{\alpha}{\theta} = \frac{d_1}{D} \text{ (p. 156) and } Q = \frac{T}{\pi}\frac{C.d_1}{2D}$$

for each instrument.

Determine the time T by the method of Section III and calculate the value of the instrument constant $\frac{T}{\pi}\frac{C}{D}$.

We have deduced the value of α under the condition that the needle or coil moves without damping, which is never quite the case. But the damping may be taken into account by multiplying the observed amplitude by $e^{\lambda/2}$, since the amplitude of an oscillating system is reduced in the time t from 1 to $e^{-2\lambda t/T}$, p. 273, and the time taken up by the needle

* See Starling, *Electricity and Magnetism*, Chap. IX.

or coil in moving from its position of rest to its first greatest elongation is $T/4$*.

The final equations for Q are therefore

$$Q = \frac{T}{\pi} \frac{Ce^{\frac{\lambda}{2}} \sin \frac{\alpha}{2}}{\tan \theta}, \text{ or } Q = \frac{T}{\pi} \frac{Ce^{\frac{\lambda}{2}} . \frac{\alpha}{2}}{\theta},$$

or to a degree of approximation sufficient for the present purpose

$$Q = \frac{T}{\pi} . \frac{Cd_1}{2D} . e^{\frac{\lambda}{2}}.$$

(*b*) **Solenoid Method.**

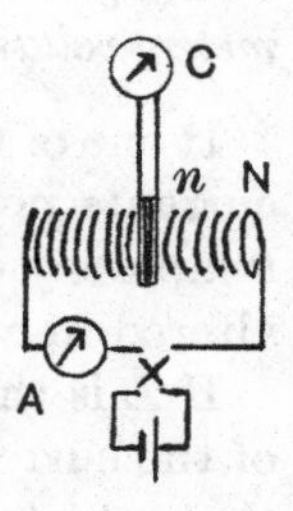

Fig. 125 *a*

If a small coil of n turns is wound round the middle of a long solenoid of N_1 turns per centimetre length and its ends connected to the galvanometer, on reversing a known current A amperes in the solenoid a quantity Q_0 of electricity is sent through the galvanometer, and if the extent of the first throw is noted we can determine the constant of the instrument.

The field in the solenoid is $4\pi AN_1/10$, and if d is the mean diameter of solenoid the magnetic flux through it is $\pi d^2/4$ times the field. On reversing this field we have

$$Q_0 = 2\frac{n}{R} . \frac{4\pi AN_1}{10} . \frac{\pi d^2}{4}.$$

Hence if d_0 is the throw of the galvanometer needle under these conditions and d_1 when a quantity Q is discharged through the instrument

$$Q = 2\frac{n}{R} . \frac{4\pi AN_1}{10} . \frac{\pi d^2}{4} . \frac{d_1}{d_0}$$

if the damping is the same in both cases.

Exercise II. Standardise the same galvanometer by this method and compare the two results. With different values of A determine to what extent the throw is proportional to Q. Plot a curve connecting the two.

* The logarithmic decrement of a moving needle galvanometer changes very little with the resistance of the circuit, but that of a moving coil instrument changes considerably and must be determined for the conditions under which it is used in the experiment.

SECTION LXVIII

THE SELF-INDUCTANCE OF A COIL

Apparatus required: *Coil of considerable inductance, Post Office resistance box, additional resistance, mirror galvanometer, voltaic cell of constant electromotive force.*

If one of the arms of a resistance bridge balanced for steady currents possesses inductance, the galvanometer will shew a temporary deflection when the current through the bridge is altered in any way.

If L is the inductance of the arm and $\dot{x}$ the rate of change of the current through it at any instant, the effect is equal to that which would be produced if an opposing electromotive force $L\dot{x}$ were introduced into the arm. Such an electromotive force would cause a current $KL\dot{x}$ to flow through the galvanometer, where the value of K depends on the resistances of the galvanometer, and of the arms of the bridge. If the current is rapidly changed from 0 to x, the electromotive force at each instant depends on the rate of change, but the total quantity of electricity which flows through the galvanometer will be KLx, which is the same as that conveyed by a constant current $KL\dot{x}$, flowing during a short time τ, such that $\dot{x}\tau = x$.

This quantity of electricity passing through the galvanometer coils will cause a deflection or "throw" of the galvanometer needle, the extent of which may be expressed in terms of the quantity and the constants of the galvanometer by the equations

$$Q = \frac{T}{\pi} \cdot \frac{C \sin \frac{\alpha}{2}}{\tan \theta} e^{\frac{\lambda}{2}} \text{ or } Q = \frac{T}{\pi} \frac{C \cdot \frac{\alpha}{2}}{\theta} \cdot e^{\frac{\lambda}{2}}$$

according to the type of galvanometer used.

To apply these equations to the determination of the inductance of a coil we substitute for Q the value KLx. Thus

$$L = \frac{1}{Kx} \cdot \frac{T}{\pi} \cdot \frac{C \sin \frac{\alpha}{2}}{\tan \theta} e^{\frac{\lambda}{2}} \text{ or } L = \frac{1}{Kx} \cdot \frac{T}{\pi} \frac{C \cdot \frac{\alpha}{2}}{\theta} e^{\frac{\lambda}{2}}.$$

The quantities K and C are eliminated in the following way. If the balance of the bridge be disturbed by inserting a small resistance r in series with the coil, this additional resistance will have, while a current x' is passing through it, the same effect on the galvanometer as an opposing electromotive force rx'. If this electromotive force produces the current C in the galvanometer, then $C = Krx'$ (p. 334). Hence

$$L = r \frac{x'}{x} \cdot \frac{T}{\pi} \cdot \frac{e^{\frac{\lambda}{2}} \sin \frac{\alpha}{2}}{\tan \theta} \text{ or } L = r \cdot \frac{x'}{x} \cdot \frac{T}{\pi} \frac{e^{\frac{\lambda}{2}} \cdot \frac{\alpha}{2}}{\theta}.$$

In most cases the calculation may be further simplified, since x'/x is the ratio of the currents in the coil with and without the additional resistance r, which should always be so small that x' is sensibly equal to x.

If α and θ are sufficiently small these angles may be taken to be proportional to the observed deflections d_1 and D, otherwise $\sin \frac{\alpha}{2}$ and $\tan \theta$ must be evaluated according to the equations given on p. 156. But it follows from these equations that the ratio $\sin \alpha/2 \big/ \tan \theta$ is equal to $d_1/2D$ to the second approximation if

$$\frac{D}{d_1} = \sqrt{\frac{11}{8}} = 1{\cdot}173,$$

and that $\frac{\alpha}{2\theta} = \frac{d_1}{2D}$ to the second approximation if $D = d_1$. Either may be secured by a proper choice of r.

Care should be taken to have the zero reading of the galvanometer near the centre of the scale and to take deflections towards both sides, and thus eliminate any errors due to faulty adjustment of the scale.

The principal experimental difficulty in the above determination of the inductance of a coil will be found to be due to the gradual heating of the coil by the current, which must be sufficiently strong to give a measurable throw on making or breaking the circuit. Subject to this condition, the current through the coil should be as weak as possible. It is important therefore that the arms of the bridge should be arranged so that the galvanometer is most sensitive to variations of resistance in one arm of the bridge, *for a given current* in that arm. This condition leads to results which are quite different from those which apply when the electromotive force is given (p. 277). The best arrangement of the bridge when S is the resistance whose inductance is to be determined is that in which the resistance we have denoted by Q in the previous exercises is as small as possible, and that denoted by R as large as possible*. One of the galvanometer terminals should be connected to the junction of P and Q, the other to the junction of R and S. The best resistance of the galvanometer of a given type is

$$G = \frac{P(S+Q)}{P+Q},$$

or, when P is large compared to Q, $G = S + Q$, S being the resistance of the coil. The sensitiveness is not much reduced if the resistance of the galvanometer lies between $\frac{1}{4}$ and 4 times the best value.

The change of resistance r may be obtained by connecting a high resistance W in parallel with the coil, when the steady deflection θ is taken. r is then equal to $S^2/(S+W)$. The shunt must be disconnected when the throw is taken. We may vary the resistance of R instead of S, remembering that a change r in the resistance of S is equivalent to a change Pr/Q in the resistance R. If r' is the change in R which produces a steady deflection θ, we must therefore substitute for r the quantity $r'Q/P$. This is in general the more convenient way of disturbing the balance of the bridge, as it is often necessary in order to get a perfect balance to adjust R to less than an ohm by placing a resistance in parallel with a portion of it. This auxiliary resistance then serves to make the required change of R.

To carry out the experiment, arrange the given coil S to

* See Schuster, *Phil. Mag.* p. 175 (1894).

form the fourth arm of the Post Office bridge; Q being the ratio arm adjoining S. Make $Q = 10$ and $P = 1000$. Connect the galvanometer through a commutator to the junction between P and Q, and to that between R and S. Introduce also a commutator into the battery circuit, so that the current may easily be reversed. It is important that the battery should be one giving a constant electromotive force, *e.g.* a Daniell or storage cell. Leclanché cells are too variable. Adjust the zero of the galvanometer to the centre of the scale.

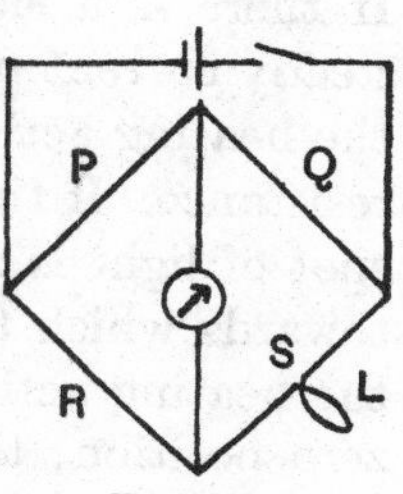

FIG. 126

Leave the galvanometer circuit open, pass the current through the coil and notice whether the galvanometer shews a deflection. If it does, the coil acts directly on the needle and its position should be changed so that the action ceases. This is most easily secured by placing the coil with its plane nearly horizontal and tilting it till the magnetic field due to it is vertical at the galvanometer. Now close the galvanometer circuit and balance for steady currents.

When the balance is approximately made, break the battery circuit and take an observation of the throw, which should be between 10 and 20 cms. as measured on the scale. If the throw is too large, the electromotive force of the battery should be diminished or a resistance inserted in the battery branch. If it is too small, the galvanometer should be made more sensitive, and if that is not possible the electromotive force of the battery must be increased. Having obtained an approximate value of the throw, the change of resistance r' in the arm R, which gives a steady deflection of between 1·1 and 1·2 times that of the throw, should be found.

If the galvanometer has a small logarithmic decrement and takes a long time to come to rest a small coil in series with an auxiliary battery and key may be fixed behind it, and connection may be made and broken in such a way as to bring the suspended system to rest.

The above preliminary experiments having been concluded, note the position of rest of the galvanometer needle. Obtain as accurate a balance of the bridge as possible, if necessary using an auxiliary resistance as a shunt in a portion of R. If there is a slow creeping of the spot of light while the steady current passes through the coil, it is probably due to the heating action of the current, which slowly increases its resistance. If that is the case, adjust the balance so that the spot of light is deflected a little to the side *opposite* to that towards which the creeping takes place. Wait till owing to the heating action the spot of light occupies accurately the zero position; then suddenly break the battery circuit and observe the throw, reading both the first deflection (d_1) and that which follows it *on the same side* (d_2). Make the battery circuit again, change the resistance R by the addition of r' and observe the steady deflection. Break the galvanometer and battery circuits and read the zero.

Reverse the commutator in the battery circuit and repeat the observations, both for the throw and the steady deflection.

Reverse the galvanometer circuit and again repeat the observations, first with the battery commutator in one and then in the other position.

Four sets of readings are thus obtained the means of which must be combined in the final result.

Determine the time of oscillation of the galvanometer needle by observing the time of say 50 swings if the time be short and the logarithmic decrement small; or if the time be long, observe the instant of passage through zero for say six swings, allowing an interval of six, then observe six more and combine in the usual way.

If it is desired to determine the logarithmic decrement this should now be done by one of the methods explained above, care being taken that the sensitiveness of the galvanometer is the same as that used during the experiment.

Tabulate your results as follows:

Date:

Inductance of coil A. Galvanometer No. 327.

$P = 1000$. $Q = 10$. 2 Daniell Cells.

Balance obtained with $R = 2194$.

On breaking circuit

First swing to right (mean of four observations) = 11·04 cms.

Second „ „ „ „ = 7·56 „

Changed R to 2255; $r' = 61$, $\therefore r =$ ·61.

Steady deflection (mean of four observations) = 12·11 „

Time of oscillation = 3·44 secs.

$$\therefore L = \cdot 61 \frac{3\cdot 44}{2\pi} \cdot \frac{11\cdot 04}{12\cdot 11} \cdot \left(\frac{11\cdot 04}{7\cdot 56}\right)^{\frac{1}{4}}$$

$= \cdot 335$ henries.

SECTION LXIX

COMPARISON OF SELF AND MUTUAL INDUCTANCES

Apparatus required: *Two coils of several hundred turns, 2 resistance boxes, cell and mirror galvanometer.*

The mutual inductance M between two coils may be determined in terms of the self-inductance L of one of them by arranging that the effect of the self-inductance on the galvanometer of a balanced resistance bridge when the current through the bridge is started or stopped is neutralised by the effect of the mutual inductance between the first coil and a second placed in the cell circuit.

The arrangement is shewn in Fig. 126 *a*. S is the coil whose self-inductance is supposed known. The resistances PQR are from the usual form of resistance box and are supplemented by a second box V, U, through which the second coil is connected to the bridge. The cell is shewn connected to an intermediate point of this resistance, but in practice it is connected to one end or the other, that is, either $U = 0$ or $V = 0$.

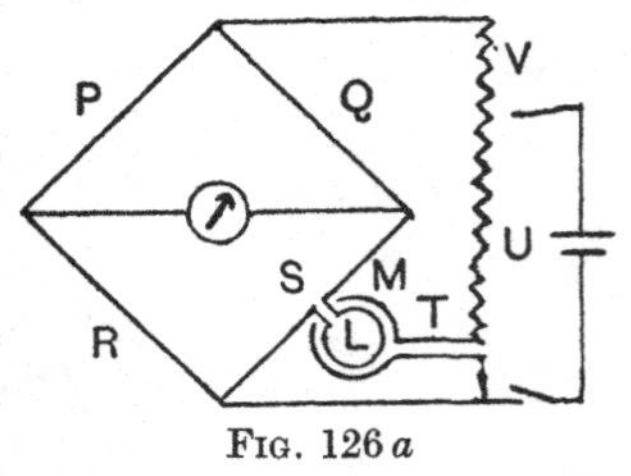

FIG. 126 *a*

For a steady current balance we have $R/S = P/Q$.

For an aggregate or integral balance for changes from one steady state to another we have

$$\frac{L}{M} = -\frac{Q + S + \left(1 + \dfrac{Q}{P}\right) V}{T + U}*.$$

The condition for a continuous balance for all changes of E.M.F. on the bridge, *i.e.* $LL' - M^2 = 0$, is not satisfied unless there is no magnetic leakage between the coils. Thus the

* Lees, *Proc. Phys. Soc.* XXVIII, p. 91 (1916).

balance is in general an aggregate or integral one only. In most practical cases an aggregate balance introduces no appreciable unsteadiness of the needle.

In the practice of the method P and Q are the proportional arms of a resistance bridge box, such as the Post Office box, and should be of the same order of resistance as that of the coil whose self-inductance is to be compared, but need not be equal. The coil of resistance S is placed in the open arm of the bridge. In boxes in which the galvanometer leads are connected to the ends of P and Q the final adjustment of the third arm R to give a balance for steady currents may be made by inserting between P and R a short length of resistance wire and altering the length till R is increased by the necessary fraction of an ohm. The actual resistance of the wire may be ascertained by observing the deflections of the galvanometer for the nearest integral ohm too small and that too great and taking proportional parts. If the connections of the keys of the box do not allow a wire to be inserted in this place, a corresponding short length may be put in series with the coil S and its value calculated by taking deflections in the same way. The second coil should be connected through a second resistance box VU to those terminals of the bridge to which the cell is usually attached, and the cell in the first instance connected to the same two terminals. The steady balance is then obtained by taking out the infinity plug at U, making the cell circuit before the galvanometer circuit, and adjusting R by means of the plugs and the wire till the deflection of the galvanometer is reduced to zero. If the drift of the value of R for a balance is considerable, the heating of the coil S owing to the current in the bridge must be reduced by inserting a resistance in the cell circuit.

The trouble due to this drift is least when the inductive balance is obtained on breaking the cell circuit rather than on making. The direction and magnitude of the throw of the galvanometer on breaking the cell circuit should be noticed. This direction is that of the self-inductance throw. Now insert the infinity plug in U so that the mutual inductance comes into play, and again notice the direction and magnitude of the throw. If the second or composite throw is greater algebraically than the first or self-inductance throw, the second or mutual inductance coil must have its connections to the bridge reversed. If the composite is less algebraically than the self throw, the connections are right. If the composite throw is zero we have a direct balance. If the composite throw is of opposite sign to the self throw the resistance U should be increased till the composite throw becomes zero, V being kept zero. If the composite throw is in the same direction as, but less than, the self, transfer the cell lead to the U end of the resistance box, thus making U zero, and adjust V till the throw is zero. None of these alternatives requires the original steady balance to be disturbed, and by following the routine suggested there is no time wasted in useless observations, whatever the quotient of L by M may be.

Record as follows:

Date:

Coils A $(= L)$ and B.

Coil A in bridge arm. $P = Q = 100$ ohms.

Steady balance with $V = 0$, $U = \infty$. With $R = 77$ deflection 17·7 left, $R = 76$, 16·1 right. Hence R for balance = 76·48 ohms. Wire resistance adjusted to give steady balance.

Induction balance. Throws on breaking cell circuit.

With $V = 0$, $U = \infty$, self 25 left; $U = 0$, composite off scale left.

Connections of mutual coil reversed; composite 4·3 left.

Hence U made $= 0$ and V adjusted till with $V = 22$ ohms composite throw zero. T found 129·7 ohms.

Hence $L/M = -220{\cdot}48/129{\cdot}7 = -1{\cdot}70$.

SECTION LXX

LEAKAGE AND ABSORPTION IN CONDENSERS

Apparatus required: *Condensers, one of which is specially selected to shew leakage and absorption, high resistance galvanometer, discharge key, cells.*

Unless the plates and terminals of a condenser are well insulated, the condenser is found gradually to lose its charge. The dielectric itself may not be a sufficiently good insulator. The rate at which the loss takes place is generally greater the greater the charge which the condenser possesses.

To determine the extent to which leakage takes place in the dielectric of a paper condenser C, connect it through a "condenser key" K to one or two Leclanché cells L and a galvanometer G, as shewn in the figure. The whole of the circuit must be well insulated.

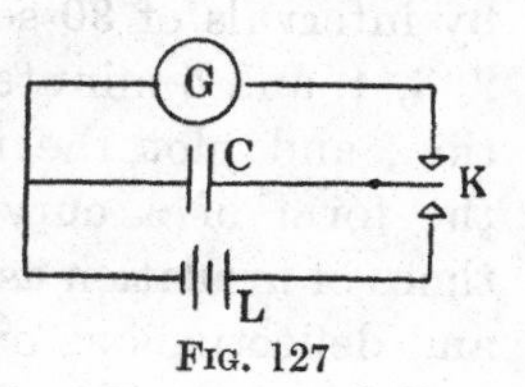

FIG. 127

When the moveable part of the key is down, the condenser is in connection with the cells and is charging; when the disc marked "insulate" on the key is pressed, the moving arm of the key takes up the position shewn in Fig. 127, and terminals of the condenser are insulated from each other; when the disc marked "discharge" is pressed, the arm moves up to the upper contact in the figure, the terminals are connected together through the galvanometer, and the condenser discharges itself.

If the capacitance of the condenser is C, the charge Q which an electromotive force E imparts to it is equal to EC.

Owing to the passage of the discharge through the galvanometer the needle is deflected and the extent of the first swing a from the position of rest is proportional, within the degree of accuracy required for the present purpose, to the amount Q of electricity discharged through the galvanometer.

Take several observations of the swings when charging and discharging are performed in the least possible time. In order that the condenser may be properly discharged between each observation, allow the key to remain at the discharge position for three minutes before again charging. The mean of these deflections may be taken to represent the charge of the condenser due to the applied electromotive force before leakage has had time to occur. They ought not to be less than 20 large scale divisions.

Now take several observations, charging the condenser for an instant only as before, but insulating for ten seconds, then discharging, and allowing three minutes before again charging.

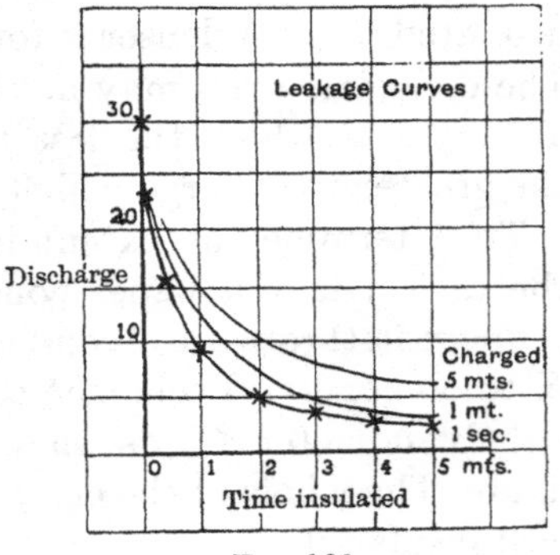

FIG. 128

Repeat the observations for instantaneous charges followed by intervals of 30 seconds, 1, 2, 3, 4, and 5 minutes' insulation, and plot the results in the form of a curve, taking times of insulation as abscissae and deflections as ordinates.

Take logarithms of the deflections, subtract consecutive logarithms from each other, divide by the differences in the time of insulation, and tabulate the results as follows:

Condenser A. Leakage.

Time of insulation	Deflection on discharge	Log. of deflection	Difference	Diff. of logs. / Diff. of times
0 sec.	30·9	1·490		
10 „	23·3	1·367	·123	·0123
30 „	15·8	1·199	·168	·0084
1 min.	9·0	·954	·245	·0082
2 „	5·5	·740	·214	·0036
3 „	3·6	·556	·184	·0031
4 „	2·9	·470	·086	·0014
5 „	2·4	·380	·090	·0015

If the leakage at any instant were proportional to the charge in the condenser at that instant, the numbers in the last column would be equal. Their gradual diminution shews that the leakage increases more rapidly than the charge.

In order to test whether the inequality is due to absorption of the charge by the dielectric of the condenser, the effect of variation in the time of charge should now be found. The condenser should be charged each time for 10 seconds instead of an instant only, and the rest of the experiment carried out as previously. Then the charging should be continued for 30 seconds and afterwards for 1 and for 5 minutes.

Tables of results for each time of charge should be given and leakage curves drawn alongside the one drawn previously.

If absorption by the dielectric occurs, the absorbed charge should make its reappearance after discharge on the insulated conductors of the condenser. To test this, if the condenser has no short circuiting key, arrange one in parallel with the condenser. Charge the condenser for 10 seconds, discharge for an instant through the short circuiting key and then insulate for 5 seconds. On now pressing the discharge key, the galvanometer will be deflected owing to the passage through it of the "residual" discharge.

Repeat the observations, allowing the condenser to remain insulated 10, 20, 40, and 60 seconds before taking the residual discharge.

To find the influence of the time of charging on the magnitudes of the residual discharges, charge the condenser for 20, 40, and 60 seconds, insulate in each case for 5, 10, 20, 40, 60 seconds and take the first residual discharges as before.

Express the results by curves and in tabular form, as follows:

Condenser A. First Residuals after charge for given time, discharge, and subsequent insulation for given time.

Time of charge	First residuals after insulation for				
	5 secs.	10 secs.	20 secs.	40 secs.	60 secs.
5 seconds	8·0	9·0	10	11	12
10 ,,	10	11	12	14	17
20 ,,	11	13	14	17	20
40 ,,	13	16	17	20	23
60 ,,	14	18	20	23	26

To test for successive residuals, charge again for 20 seconds, then discharge through the short circuiting key. Insulate for 10 seconds, take the first residual, insulate for 10 seconds more and take the second residual, and so on till no more charge remains in the condenser.

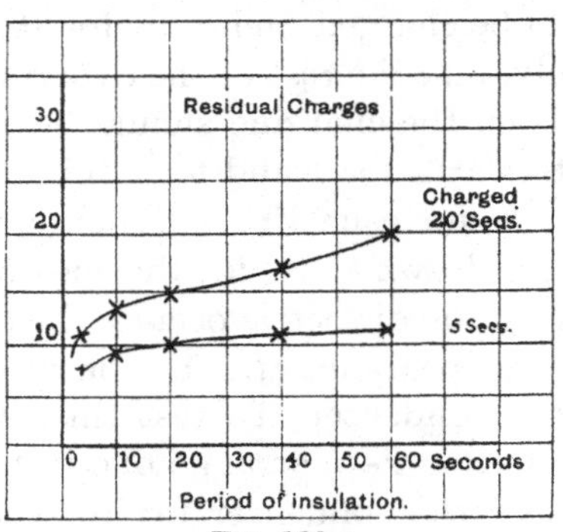

FIG. 128 *a*

If the galvanometer is not sufficiently damped to allow of the observations to be taken so rapidly after each other, take the residuals after intervals of 20 or more seconds. If the galvanometer takes much longer to come to rest, after taking the first residual in the way described, recharge the condenser, discharge, take the first residual through the short circuiting key, and the second residual through the galvanometer. Similarly for successive residuals.

Tabulate as follows and draw a curve with the magnitudes of the successive residuals as ordinates, and times of insulation since the first discharge, as abscissae.

After insulation for	Residual discharges
10 seconds	13·8
10 seconds more	7·2
10 ,, ,,	6·4
10 ,, ,,	5·2
10 ,, ,,	4·1

Take a second condenser not specially chosen to exhibit leakage and absorption, and test whether the leakage and absorption are small, as they ought to be in a good condenser.

SECTION LXXI

COMPARISON OF CONDENSERS

Apparatus required: *Condensers, discharge key, high resistance mirror galvanometer, cells.*

(*a*) **Ballistic Methods.**

It has been stated in the previous section that when a condenser is discharged through a galvanometer, the extent of the first swing from rest is proportional to the quantity of electricity discharged. If Q is this quantity, E the potential of the battery used to charge the condenser, and C the capacitance of the condenser, then $Q = CE$.

The method of discharge may therefore be used either to compare the capacitances of a number of condensers charged by the same battery and discharged through the galvanometer, or to compare the electromotive forces of a number of cells used to charge the same condenser. It is seldom used for the latter purpose as the potentiometer method is more accurate.

Apply the method, using a Leclanché cell and a high resistance galvanometer which swings without too much damping, to determine the capacitances of the condensers provided, assuming that of the standard condenser to be given in microfarads. The condensers should be placed in circuit in turn, charged for an instant, then discharged, and the extent of the swing of the galvanometer needle observed.

In order to see whether leakage has any effect on the observations, take discharges from each condenser after 10 and 20 seconds' insulation, as well as instantaneous discharges.

Record as follows:

Date:

Nalder Galvanometer No. 5232. Leclanché Cell No. 12.

Condenser	Discharge			Capacitance microfarads	Remarks
	Instantaneous	after 10 secs.	after 20 secs.		
Standard	23·7	23·8	23·7	·33	no leak
Swiss	26·4	25·2	25·0	·37	small leak
Paraffin No. I	27·2	26·1	25·4	·38	,, ,,
,, ,, II		&c.			
,, ,, III					

(*b*) **Balance Methods.**

1. *Gott's method.*

Two condensers, C_1, C_2, are connected in series with each other to 3 or 4 Leclanché cells in series (Fig. 129). Two resistances, R_1, R_2, in series are connected to the cells. The galvanometer connects the joining point of the two condensers with that of the two resistances. One of the resistances is varied till on making or breaking the cell circuit at the key K there is no throw of the galvanometer needle. Then

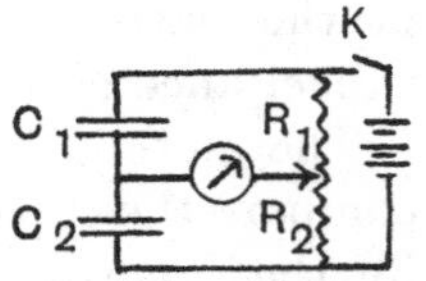

Fig. 129

$$R_1 C_1 = R_2 C_2.$$

As a rule the leakage and absorption of the condensers prevent the balance being perfect. A slight throw is generally observed, followed by a slow drift. The resistance should be varied till the throw is reduced to a minimum.

2. *De Sauty's method.*

The cells and galvanometer are interchanged. The condition for a balance is unchanged.

Compare two of the condensers by these methods.

Record as follows:

Condenser	Resistance		
	Gott	De Sauty	Mean
Standard C_1	1000	1000	1000
Swiss C_2	905	907	906

$$\therefore C_2 = 1{\cdot}12\, C_1.$$

(*c*) **Alternating Current Methods.**

If in either of the balance methods the Leclanché cells are replaced by a source of alternating current and the galvanometer is replaced by a telephone, absence of sound in the telephone implies $R_1C_1 = R_2C_2$. Leakage and absorption prevent a perfect balance and a minimum only is obtained.

Compare the two condensers using the alternating current from a small induction coil as on p. 307.

SECTION LXXII

THE CAPACITANCE OF A CONDENSER

Apparatus required: *Condenser, high resistance galvanometer, resistance coils, voltaic cell.*

When Q coulombs of electricity are discharged through a galvanometer the needle is deflected and swings through an angle α which, neglecting the damping, has been proved (pp. 331–2) to be given by the equations

$$Q = \frac{H}{G} \cdot \frac{T}{\pi} \sin\frac{\alpha}{2} \quad \text{or} \quad Q = \frac{N}{AH'} \frac{T}{\pi} \cdot \frac{\alpha}{2},$$

where H/G is the constant of the suspended needle, and N/AH' that of the suspended coil galvanometer with radial field.

If the quantity Q is the charge of a condenser of capacity C, due to an applied electromotive force E, then $Q = CE$ and therefore

$$C = \frac{1}{E} \cdot \frac{H}{G} \cdot \frac{T}{\pi} \sin\frac{\alpha}{2} \quad \text{or} \quad C = \frac{1}{E} \cdot \frac{N}{AH'} \frac{T}{\pi} \cdot \frac{\alpha}{2}.$$

The damping may be taken into account by introducing the factor $e^{\lambda/2}$ on the right-hand side. The quantity $\frac{1}{E} \cdot \frac{H}{G}$ or $\frac{1}{E} \cdot \frac{N}{AH'}$ may be determined or eliminated by observing the steady deflection of the galvanometer produced by the electromotive force E acting through known resistances.

Let a circuit be arranged as in Fig. 130, the cell which has served to charge the condenser having its terminals connected to a resistance R, which should be large compared to that of the cell, so that the difference of potential at the ends of R may be taken to be equal to the electromotive force E. The terminals of the galvanometer are connected only to a small portion r of R, so that the difference of potential at the ends of the galvanometer circuit is Er/R. If the resistance of the galvanometer be ρ and the resistance ρ' shewn in the figure be not inserted, the current passing through the galvanometer will be $Er/\rho R$, and the permanent deflection will be given by the equations

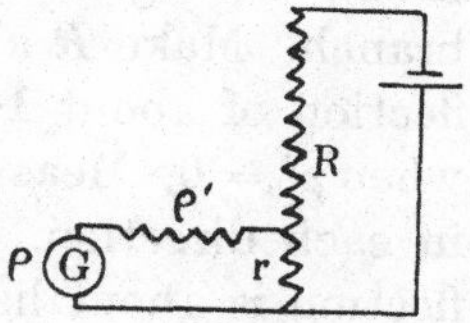

Fig. 130

$$\frac{Er}{\rho R} = \frac{H}{G} \tan\theta \quad \text{or} \quad = \frac{N}{AH'}\theta;$$

$$\therefore C = \frac{r}{R} \cdot \frac{T}{\pi} \cdot \frac{e^{\lambda/2} \sin\frac{\alpha}{2}}{\rho \tan\theta} \quad \text{or} \quad = \frac{r}{R} \cdot \frac{T}{\pi} \cdot \frac{e^{\lambda/2}\frac{\alpha}{2}}{\rho \, . \, \theta}.$$

If the galvanometer resistance ρ is unknown it may be obtained, or eliminated from the above expression, if a known resistance ρ' is inserted in the galvanometer circuit and the steady deflection θ' observed. We have in that case

$$\frac{Er}{(\rho + \rho')R} = \frac{H}{G} \tan\theta' \quad \text{or} \quad = \frac{N}{AH'}\theta';$$

and combining this with the equation obtained for the case when ρ' is not inserted, we find

$$\rho = \frac{\rho' \tan\theta'}{\tan\theta - \tan\theta'} \quad \text{or} \quad = \frac{\rho'\,\theta'}{\theta - \theta'}.$$

The condition most favourable for the determination of ρ by this method is that in which ρ and ρ' are equal, so that the introduction of ρ' approximately halves the deflection.

The remarks made on p. 336 as to the best way of determining $\dfrac{e^{\lambda/2} \sin\frac{\alpha}{2}}{\tan\theta}$ or $\dfrac{e^{\lambda/2}\frac{\alpha}{2}}{\theta}$ apply in the present case. λ should be determined under the conditions of the experiment.

Determine the capacity of the condenser provided, charging it by two Leclanché cells and discharging it through a well-insulated circuit including the galvanometer. Take four observations of the first throw, two of which should be made with the battery reversed.

Connect the galvanometer to the two Leclanché cells as shewn in Fig. 130, adding a commutator in the battery branch. Make R about 10,000 ohms, and r such that a deflection of about 1·1 times the discharge throw is obtained when $\rho' = 0$. Measure the deflection with the battery current in each direction. Insert a resistance ρ' such that the deflection is about half what it was in the previous case, and again make two determinations of the deflection, reversing the battery current between the two observations.

Measure the time of swing and the logarithmic decrement, following the instructions given on p. 273.

Record your observations as follows:

Date:

Standard Condenser marked 1/3 microfarad.

Galvanometer No. 5232.

1. Discharge of Condenser after being charged by two Leclanché Cells.

First deflection (mean of two deflections in each direction) = 16·15 cms.

2. Measurement of steady Deflection.

$$R = 10{,}000,\ r = 40.$$

(*a*) $\rho' = 0$. Deflection (mean of one observation in each direction) = 18·35 ,,

(*b*) $\rho' = 6170$. Deflection (do.) = 9·45 ,,

Hence galvanometer resistance

$$\rho = 6170 \frac{9{\cdot}45}{18{\cdot}35 - 9{\cdot}45} = 6580.$$

A previous direct measurement of the galvanometer resistance had given $\rho = 6570$. Either value may be adopted in the subsequent calculation.

3. Determination of time of oscillation.

Time of 20 swings (average of 3 observations) = 74·9 secs.

,, ,, one swing = 3·745 ,,

4. Determination of logarithmic decrement.
Zero 0.

	Successive turning points on right	on left	
(1)	19·58	– 16·16	scale divisions
(2)	12·37	– 10·33	,,
(3)	8·34	– 6·80	,,
(4)	5·41	– 4·61	,,

Ratio of (3) : (1) = 2·38 and 2·35
,, ,, (4) : (2) = 2·24 ,, 2·29

Mean Ratio = 2·315
Mean of three determinations = 2·310

$$\lambda = \frac{1}{4}\frac{\log 2{\cdot}31}{\log e} = \frac{1}{4} \times \frac{\cdot 3636}{\cdot 4343} = \cdot 2092.$$

$$1 + \frac{\lambda}{2} = 1{\cdot}105.$$

5. Calculation of Capacitance.

$$C = \frac{40 \times 3{\cdot}745 \times 1{\cdot}105 \times 16{\cdot}15}{10000 \times 2\pi \times 6580 \times 18{\cdot}35}$$

$$= \cdot 337 \times 10^{-6} \text{ farads}$$

$$= \cdot 337 \text{ microfarad.}$$

SECTION LXXIII

HIGH RESISTANCE BY CONDENSER

Apparatus required: *Condenser of about $\frac{1}{3}$ microfarad, resistance of about* 10^9 *ohms, and ballistic galvanometer.*

When a resistance R of the order of 10^9 ohms is to be measured the methods described in Section LIV are no longer suitable. It can however be determined by connecting it to the terminals of a charged condenser of known capacity C and measuring by a ballistic galvanometer what fraction of the charge remains in the condenser after a definite time. If Q_0 is the original charge and Q its value after it has leaked for t seconds, then

$$\frac{1}{R} = \frac{C}{t} \log_e \frac{Q_0}{Q}.$$

If C is expressed in farads the resistance is given in ohms.

Exercise. Determine the resistance of a short rod of fibre by this method.

Connect the condenser to the ballistic galvanometer and cell as shewn in Fig. 131 and connect one end of the high resistance R to one terminal of the condenser permanently and the other end to the other terminal through a key. With the key out determine the throw of the galvanometer on charging and discharging the condenser instantaneously. Charge again for an instant and insulate for about 2 minutes. At an observed time discharge again and note the first throw.

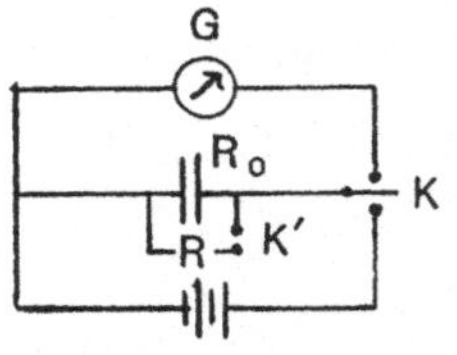

FIG. 131

Repeat two or three times and take the means. Calculate from them by means of the expression given the resistance R_0 of the condenser and connections.

Connect the unknown high resistance across the terminals of the condenser and repeat the observations. From the means in this case calculate the resistance R_1 of R_0 and R in parallel. Then since $\frac{1}{R_1} = \frac{1}{R_0} + \frac{1}{R}$ R may be calculated.

Record as follows:

Date:

Condenser $\frac{1}{3}$ microfarad standard.

Ballistic galvanometer C.

High resistance fibre rod.

Electromotive force 2 Leclanché cells.

Without resistance.

Q_0 gave deflection 21·0, 21·1, 20·9, mean 21·0.

Q after 100 seconds 19·1, 19·1, 19·0, mean 19·1.

$$\text{Hence} \quad \frac{1}{R_0} = \frac{\cdot 333 \times 10^{-6}}{100} \log_e \frac{21\cdot 0}{19\cdot 1} = 3\cdot 17 \times 10^{-10}.$$

Similarly with the resistance connected to the condenser.

SECTION LXXIV

THE CHARACTERISTIC CURVES OF A TRIODE TUBE

Apparatus required: *Triode tube, three milliammeters, adjustable resistance, storage cells.*

The Triode tube or Wireless Valve is a vacuum tube containing (1) a metal filament F, generally tungsten, through which a current can be sent which raises its temperature and causes it to emit electrons at a rate which increases rapidly as the temperature rises; (2) an open coil of wire G, generally nickel, wound round the filament F at about 5 mms. from it, the turns of the coil being about 1 mm. apart; (3) a plate P, generally nickel and cylindrical in form, surrounding the coil.

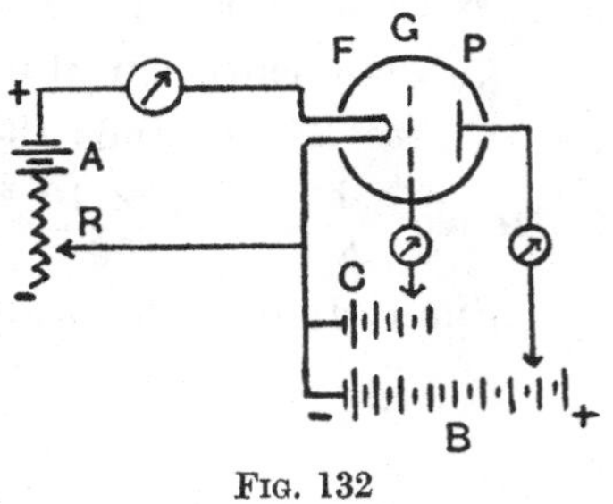

FIG. 132

A few storage cells A provide the current for heating up the filament. It is adjusted by the resistance R and read on the ammeter. This current should never exceed that for which the filament is intended.

If an electromotive force be applied by means of the cells B between the filament and plate so that the plate is positive with respect to the filament, the electrons produced by the filament will move towards the plate with a speed which depends on the acceleration the electric field produces. If the plate is negative with respect to the filament the opposite effect is produced.

If an E.M.F. is applied between the filament and grid G the electrons moving towards the plate will be accelerated if the grid is positive and retarded if it is negative. Since the grid is nearer to the filament than the plate a small change of the potential of the grid has a much larger effect on the current between filament and plate than it would have if applied to the plate. The grid thus provides a means of altering the current in the plate-filament circuit without any change of the electromotive force applied to that circuit. As the current in the filament-grid circuit is small the energy which has to be supplied to the grid circuit to produce a large change of current in the plate circuit is small.

To determine the characteristic curves:

(1) *The grid not in circuit.*

Connect cells A to the filament and send a current of ·1 ampere through it. Apply between the filament and plate electromotive forces varying from − 100 to + 200 volts by means of the cells B or the laboratory mains, observing the

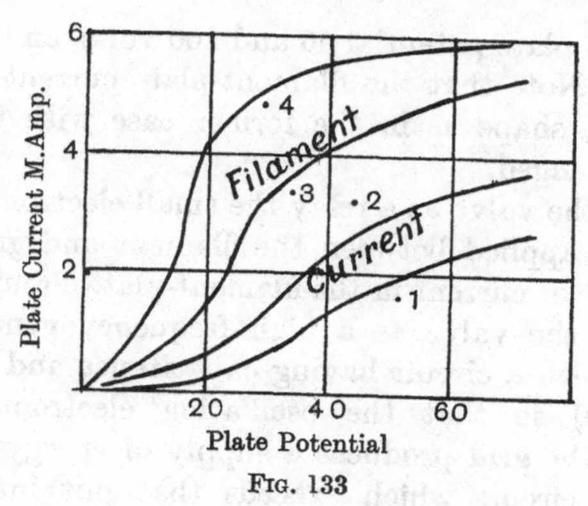

Fig. 133

volts applied and the current which passes. Plot a curve with potential applied as abscissae and current in milliamperes as ordinates. Increase the filament current to ·2, ·3, ·4 ampere up to the maximum the filament is designed to take. In each case plot the characteristic curve. Notice that the curves are all of the same shape.

(2) *The grid in use.*

Adjust the potential on the plate to 100 volts and the filament current to ·4 ampere. Apply between the filament and grid an electromotive force varying from − 50 to + 50 volts by means of cells or laboratory mains. Observe at each E.M.F. the currents in the plate and the grid circuits, and plot the results thus:

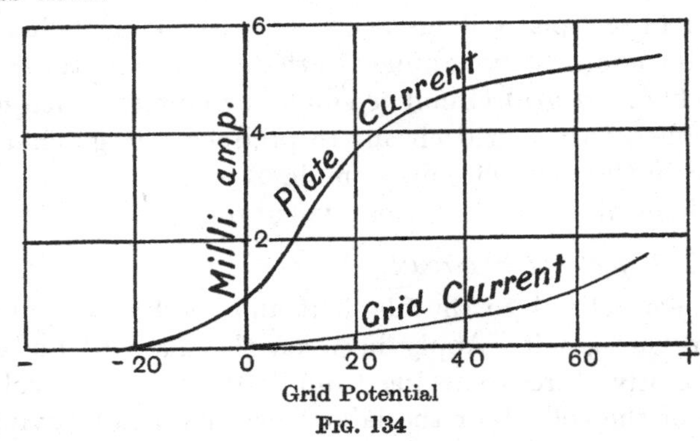

FIG. 134

Repeat with plate potential 50 and 200 volts, and plot in the same way. Note that the filament-plate current curves are of the same shape as in the former case with the zero of abscissae changed.

In using the valve as a relay the small electromotive force available is applied between the filament and grid and the large change of current in the filament-plate circuit obtained.

In using the valve as a high frequency generator it is combined with a circuit having capacitance and inductance (capindance) so that the oscillating electromotive force applied to the grid produces a supply of energy in the filament-plate circuit which exceeds that put into the grid circuit, and the connections are so arranged that the system becomes unstable and small oscillations in the circuit are magnified into large, the frequency remaining that due to the capacitance and inductance of the circuit*.

* See Fortescue, *Dict. App. Physics*, vol. II, p. 882.

SECTION LXXV

THE QUADRANT ELECTROMETER

Apparatus required: *A Dolezalek Electrometer and scale, key, water resistance and condenser.*

The quadrant electrometer consists of a light moving needle N, Fig. 135, enclosed in a fixed cylindrical box cut into four quadrants, Q_1, Q_2. Opposite quadrants are connected together, and the needle is generally at a potential differing by 50 to 200 volts from that of the quadrants. The metal case C containing the working parts of the instrument is earthed.

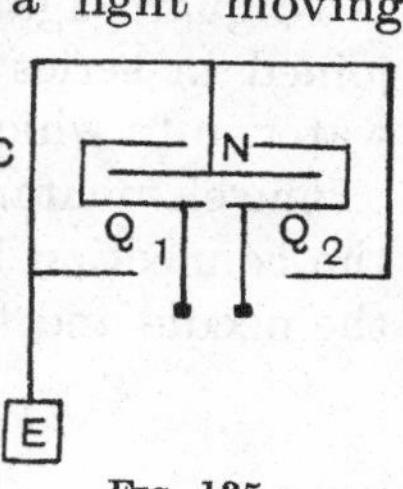

FIG. 135

Adjustments. The needle should first be arranged with its long axis inclined at 5–7° to one of the slits dividing the quadrants. This can be done by a screw head at the top of the electrometer.

The needle should turn in a plane parallel to the plane of the quadrants. This can be secured by levelling the electrometer by means of the three levelling screws. Remove the case of the instrument and adjust the screws till, when looking along either of the slits between the quadrants, the suspension is seen to be symmetrical. Replace the case and with quadrants and needle earthed read the position on the scale. Charge the needle, and if the position on the scale has altered, bring it back by turning one of the screws slightly. Repeat the process till the motion of the spot on charging or earthing the needle is reduced to a minimum.

In order to charge or discharge the needle rapidly, make use of a paraffin block key with three holes approximately at the corners of an equilateral triangle, filled with mercury. The connecting key is a bent wire with sealing wax handle.

One pair of quadrants of the electrometer is permanently earthed. The other pair is connected to a mercury cup in a block of paraffin into which an earthed wire may be lowered by means of a string in order to earth the pair when desired.

The insulated quadrants are shielded from all external electric fields by the earthed case of the instrument. The wire connecting to the insulated mercury cup should also be surrounded by an earthed shield.

See that all connections are good as any loose contact will give trouble.

To guard against short circuits a high resistance must be joined in series with the cells used. A U-tube filled with water, into which the connecting wires dip, is sufficient.

The elementary theory of the Quadrant Electrometer gives the connection between the deflection θ, the potential V of the needle and those of the quadrants V_1, V_2 as

$$k\theta = (V_1 - V_2)\left(V - \frac{V_1 + V_2}{2}\right),$$

where k is a constant.

V_2 is usually made zero by connecting one pair of quadrants to the case and V_1 is made $= v$ the potential to be measured. The formula then becomes

$$k\theta = Vv - \tfrac{1}{2}v^2.$$

Or V is made zero by connecting the needle to the case and V_2 is made $= V_1 + v$, where v is the potential to be measured, and the formula becomes

$$k\theta = V_1 v + \tfrac{1}{2}v^2.$$

The second method has been found the more convenient.

Exercise I. Standardise the electrometer. Connect the needle through a U-tube water resistance to the case of the instrument. Connect one pair of quadrants through a water resistance to one terminal of the 100 volt circuit and the other terminal of the circuit through a water resistance to the case of the instrument. Connect the two terminals of the quadrants together and read the zero. Connect a Weston cell between the terminals of the first pair and that of the second pair of quadrants, and read the deflection. Replace the cell by 2 cells in series and read the deflection. Repeat with 3 and 4 cells. Draw a curve connecting the deflection with the volts applied.

Replace the single cell and alter the potential applied to the first pair of quadrants to 50 and to 25 volts, reading the deflection. If the light mains at 200–240 volts are available, substitute them for the 100 volts and repeat. The electromotive force of the cells and mains may be read on a standardised voltmeter. Draw a curve connecting the deflection with the volts applied.

Exercise II. Capacitance of Electrometer.

The electrometer measures primarily potential differences, and to use it for measuring quantities of electricity it is necessary to know its electrostatic capacity. The capacity of the electrometer can be measured as follows if a standard condenser is available:

Use the electrometer with the first pair of quadrants at a potential which gives a deflection of about 100 divisions when 1 volt is applied between the quadrants. Read the deflection and insulate the quadrant. Join a known capacitance in parallel with the two quadrants by means of the insulated mercury cups and note the change in deflection.

If C = capacitance of the electrometer.
C' = known capacitance.
V = potential to which electrometer is charged.
V' = resulting potential after connecting capacitance C' to the electrometer.

Then $$CV = (C + C')\,V' \quad \text{or} \quad C = C'\frac{V'}{V - V'}.$$

For the deflections observed the standardisation has shewn that they are proportional to the volts. Hence C is found in terms of C'.

The known capacitance C' consists of two round metal plates separated from each other by short pillars of ebonite. Measure the diameter of the plates and their distance apart and calculate C', using the equation

$$C' = \frac{S}{4\pi d},$$

where S is the area of the surface of a plate facing the other plate and d is their small distance apart in cms.

SECTION LXXVI

IONISATION CURRENTS BY ELECTROMETER

Apparatus required: *Ionisation chamber, uranium oxide, electrometer.*

A current of the order 10^{-10} ampere is too small to be measured by a galvanometer, but the total quantity of electricity it transfers in a few minutes can be measured by means of an electrometer.

If the current I amperes is so sent as to charge one pair of quadrants of the electrometer whose capacitance is C farads then in t seconds the potential of the pair will increase It/C volts and produce a deflection d of the instrument. If the constant of the electrometer is K then $Kd = \frac{It}{C}$ and $I = K\frac{Cd}{t}$.

Exercise. Measure the ionisation current in the air in a closed vessel produced by the presence of uranium oxide.

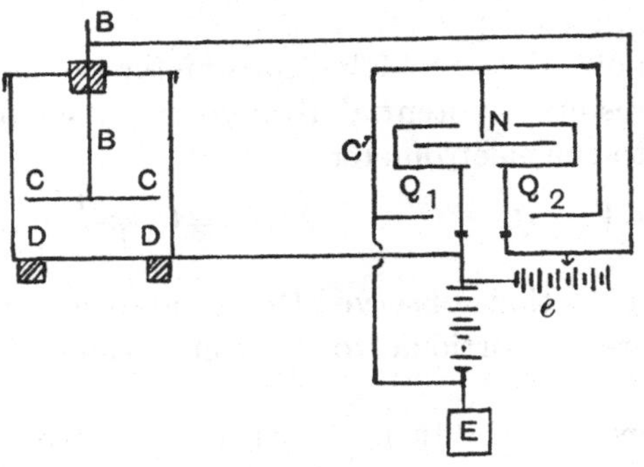

FIG. 136

The uranium oxide must be spread over the bottom of the cylinder DD, in as uniform a layer as possible. The plate CC is connected to one pair of quadrants of the electrometer, and the plate DD through the cells e driving the current to the other pair.

The whole apparatus is insulated on blocks of paraffin.

With a fixed distance l between the plates vary the electromotive force e and determine the current in each case. Plot a curve shewing the connection between the electric field and the current.

With a fixed value of e vary the distance l and observe the current. Plot as before.

From the results verify that the saturation current is proportional to l.

APPENDIX

The following details, referring chiefly to the dimensions of the apparatus used in the Physical Laboratories of the University of Manchester or the East London College, will probably prove useful to teachers.

SECTION

IV. Spirit Level.

The ordinary 8 inch "brass adjusting level" is suitable. If the tube is not graduated, a paper scale may be gummed to it.

V. Calibration Tube.

A tube about 20 cms. long, of ·7 cm. external and ·07 cm. internal diameter, may be used. The tube and mercury should be quite dry.

VI—VIII. Balance and Density.

A piece of quartz of 100–200 grams is a suitable body to weigh, since it can easily be kept clean, and therefore of constant weight. As quartz is unacted on by water it is also suitable for the density determinations.

X. Moments of Inertia.

The block should be suspended by means of a thin soft wire free from kinks.

XI. Reversible Pendulum.

A brass bar about a metre long, 2 cms. broad and ·5 cm. thick, is suitable. A number of holes about 1 cm. diameter are bored through it, most of them near one end. The bar is hung from a fixed knife edge which projects through one of the holes.

XII. Beams.

Straight-grained beams about a metre long, 1·3 to 2 cms. broad and ·7 to ·9 cm. thick, should be used.

SECTION

XIII. Rigidity.

The thinnest steel pianoforte wire (No. 30) and a wire about No. 25 are suitable. The upper end of each wire is soldered into a hole in the centre of a short length of brass wire of 1·5 to 2 mms. diameter, the lower end into a hole in a brass screw of 3 mms. diameter which screws into a hole in the axis of a cylindrical brass weight about 4 cms. diameter and 5 cms. long. The upper short wire is clamped in the support.

XIV. Viscosity.

The tubes should have a length of 50 to 80 cms., the bore should be ·5 to ·8 mm. A solution of common salt of density 1·05 to 1·10 (about 10 %) is suitable.

XV. Surface Tension.

The thin glass used for covering microscope slides is suitable for the balance method, thick capillary tube for the drop method.

XVI. Expansion of a Solid.

Tubes of about 1 cm. diameter and 60 or 70 cms. long are convenient.

XVII. Expansion of a Liquid.

The graduated stem of a broken mercury thermometer serves for the stem of the dilatometer. For water from 0° to 25° C. the bulb should have about 300 times the volume of the stem.

XVIII. Pressure Coefficient of a Gas.

The bore of the capillary tube is about ·7 mm. and the bulb about 10 cms. diameter. The distance of the fixed mark on the capillary tube from the bulb is about 20 cms. A little strong sulphuric acid should be placed in the bulb to keep the air dry. By warming the bulb slightly after a little mercury has been poured into the open tube, the volume of air enclosed may be reduced to the desired value.

SECTION

XIX. Expansion of a Gas.

A tube of ·1 cm. bore and about 25 cms. long is suitable. The divisions etched on it may be rendered more distinct by rubbing a little rouge into them. A strip of white enamelled glass should be placed between the tube and frame.

XX. Pressure and Boiling Point.

An ordinary boiling flask with the delivery tube bent upwards at the point where it joins the neck of the flask is suitable. A metal condenser or a glass condenser 40 cms. long is necessary.

XXI. Laws of Cooling.

The calorimeter of copper ·3 mm. thick is 6·5 cms. diameter and 8 cms. high. It stands on cork legs within a water jacket, the inside diameter of which is 8·5, the outside 12·5, and the height 12 cms.

XXII. Calorimetry.

A piece of rubber $1 \times 3 \times 3$ cms. is convenient.

XXIII. Specific Heat of Quartz.

The steam heater is 40 cms. long, has an internal diameter of 6·5 and an external of 9·5 cms. The internal cavity is closed at the top by a cork, and at the bottom by a metal door which can be moved aside so that the substance may be lowered into the calorimeter.

XXV. Latent Heat of Steam.

The condenser is of copper ·3 mm. thick. The box is $4 \times 3 \times 1{\cdot}5$ cms., so that it can pass freely into the calorimeter.

XXVI. Heat of Solution.

A calorimeter of thin copper 3·5 cms. diameter and 4 cms. high is used to hold the salt.

XXVII. Mechanical Equivalent of Heat.

The apparatus used is that of Puluj, with a modified arrangement for measuring the frictional couple.

SECTION

XXVIII. FREQUENCY BY THE SYREN.

It is convenient to apply part of the pressure on the top of the bellows by hand, as this enables the note of the syren to be readily varied by a slight change of the force used.

XXX. LISSAJOUS' FIGURES.

Large forks of frequencies 128 and 256 are suitable.

XXXII. SEXTANT.

The objects viewed should be at a considerable distance from the observer, and their angular distance apart should not be greater than about 20°.

XXXIII. CURVATURE OF LENSES.

The optic bench is 1 metre long and 14 cms. wide. The lens is an ordinary reading lens of 11 cms. diameter.

XXXIV. TOTAL REFLECTION.

The half cylinder has an edge of 2·0 cms.

XXXV. RESOLVING POWER OF A LENS.

The slits or spaces between the wires should be about half a millimetre apart.

XXXVI. THE PRISM SPECTROSCOPE.

The salts are contained in 2 inch corked sample bottles placed in holes in a block of wood 20 cms. by 7 cms. The crayons are the ordinary coloured crayons sold in sixpenny boxes.

XXXVIII. THE SPECTROMETER.

The model of the vernier is about 30 cms. long. The graduations are on paper gummed to wooden blocks and varnished.

XXXIX. REFRACTIVE INDEX OF A SOLID.

The small piece of mirror is provided to enable light to be reflected down on to the verniers when the readings are being taken.

SECTION

XLI. PHOTOMETRY.

For literature, standards and mean spherical candle power consult Fleming, *Journal Inst. Elect. Eng.* XXXII, p. 119 (1902).

XLIII. NEWTON'S RINGS.

A spectacle lens of ·5 diopter is suitable for C, Fig. 96 a.

XLIV. DIFFRACTION GRATING.

One of Thorp's replicas answers the purpose admirably.

XLVI. SACCHARIMETRY.

The tubes containing the sugar solution should be of glass or be glass-lined, to prevent the acid used in inverting the sugar acting on them.

There are other means of producing the half shadow field, as effective as the plate of quartz described on p. 233.

LI. COPPER VOLTAMETER.

The copper plates used are about 6 cms. square, and each has a lug about 3 cms. long, which fits into and makes good electrical contact with a spring clip.

LIII. RESISTANCE BRIDGE.

Coils of about 4, 20, 100, 500, and 2000 ohms give sufficient practice to the student.

LIV. HIGH RESISTANCES.

Strips of insulating material clamped between binding screws answer for the last portion of the exercise.

LV. LOW RESISTANCES.

A coil of about 30 cms. of No. 14 eureka wire, with connections made to it at several points of its length, is convenient.

LVII. RESISTANCE OF A CELL.

A condenser of $\frac{1}{3}$ to 1 microfarad capacity is suitable.

SECTION

LVIII. RESISTANCE STANDARDS.

The two nearly equal coils may conveniently be made of bare eureka or manganin wire wound on glass and kept in the same bath of petroleum.

LIX. TEMPERATURE CHANGE OF RESISTANCE.

About 4 metres of No. 38 copper, and 2 metres of No. 29 eureka or manganin wire wound on strips of insulating fibre, form suitable coils.

LX. RESISTANCE OF ELECTROLYTES.

A tube about 16 cms. long and ·4 cm. internal diameter enables all the solutions to be tested by means of one Post Office box of 11,000 ohms.

LXI. STANDARD CELL.

The vertical limbs of the H tube are 8 cms. long and 1·6 cms. diameter.

LXIII. POTENTIOMETER METHOD OF MEASURING CURRENTS.

The wire used for the standard low resistance should be capable of carrying the current without the increase of temperature produced altering its resistance appreciably.

LXV. EQUIVALENT OF HEAT BY THE ELECTRICAL METHOD.

The heating coil consists of 90 cms. of No. 14 platinoid wire.

LXVI. INDUCTION OF ELECTRIC CURRENTS.

The 3 inner coils each consist of one layer of 50 turns of No. 20 copper wire, and are wound in parallel. The 3 outer coils are similar but have 40 turns. Short lengths of platinoid wire are interposed between the ends of the coils and the terminals, so that whether one coil, or two in series, or three in series are in circuit the resistance is the same.

LXVII. STANDARDISATION OF BALLISTIC GALVANOMETER.

The solenoid is 50 cms. long, 3·5 cms. diameter and has 400 turns. The coil on it has 140 turns.

SECTION

LXVIII. SELF-INDUCTANCE OF A COIL.

The coil consists of 575 turns of No. 20 copper wire, the mean radius of a turn being 25·3 cms.

LXIX. SELF AND MUTUAL INDUCTANCES.

The outer coil has a mean diameter of 24 cms., the inner 18 cms. Each has three windings of 100, 200 and 400 turns. The inner coil may be set at any angle to the outer.

LXX. LEAKAGE AND ABSORPTION IN CONDENSERS.

The dielectric of the condenser used consists of paraffined paper, one side of which is shellac varnished.

LXXIV. TRIODE TUBE.

The valves sold for wireless receivers are suitable.

INDEX

Printed in the United States
By Bookmasters